DESIGNING TOILET CHAIR FOR ELDERLY PEOPLE

About the Authors

Nirmal Kaur is Senior Research Fellow in the Department of Family Resource Management Under AICRP on Home Science, Faculty of College of Home Science, G.B. Pant University of Agriculture and Technology, Pantnagar. She has published several articles in reputed journals and she has four chapters in books to her credit.

Promila Sharma is the Professor in the Department of Family Resource Management, Faculty of College of Home Science, G.B. Pant University of Agriculture and Technology, Pantnagar. She has guided several Master's and Doctoral researches. She has four books and several articles to her credit.

Gyanendra Sharma is the Professor in the Department of Agriculture Communication, Faculty of College of Agriculture, G.B. Pant University of Agriculture and Technology, Pantnagar. Many students of Agriculture Communication have been awarded Master's and Doctoral Degree under his guidance and have several articles in reputed journals.

R.C.S. Mehta Retd. Professor and Head of the Department of Production Engineering, Faculty of College of Technology, G.B. Pant University of Agriculture and Technology, Pantnagar. He has more than twenty five years of experience in teaching and guiding research related to development of agricultural tools and technologies.

DESIGNING TOILET CHAIR FOR ELDERLY PEOPLE

Nirmal Kaur, Promila Sharma,
Gyanendra Sharma and R.C.S. Mehta

CONCEPT PUBLISHING COMPANY PVT. LTD.
NEW DELHI-110059

ISBN-13 : 978-81-8069-869-9

First Published 2012

Published and Printed by

Concept
Since 1974

Concept Publishing Company Pvt. Ltd.
A/15-16, Commercial Block, Mohan Garden
New Delhi 110059 (India)
T: 25351460, 25351794, *F*: 091-11-25357109
E: publishing@conceptpub.com, W: www.conceptpub.com
Editorial Office: H-13, Bali Nagar, New Delhi-110 015, India.

Cataloging in Publication Data--*Courtesy:* D.K. Agencies (P) Ltd. <docinfo@dkagencies.com>

Designing toilet chair for elderly people / Nirmal Kaur ... [et al.].
p. cm.
Includes bibliographical references (p.) and index.
ISBN 9788180698699

1. Chair design—India—Udham Singh Nagar (District) 2. Toilets—India—Udham Singh Nagar (District) 3. Older people—Health and hygiene—India—Udham Singh Nagar (District) I. Nirmal Kaur.

DDC 749.32095451 23

Dr. B.S. Bisht
Vice-Chancellor

50 YEARS
G B Pant University of Agriculture & Technology, Pantnagar

G.B. Pant University
of Agriculture & Technology
PANTNAGAR - 263 145
Distt Udham Singh Nagar (Uttarakhand) INDIA

Foreword

Growing old is a characteristic feature of all living creatures but man takes it seriously. Old age is rear similar to childhood in many aspects. It is the most important and critical stage a human goes through. Here, the person faces problems physically, mentally, emotionally and even economically. In the changing society, India is gradually adopting a culture where the parents limit their company with their children and even children will have full freedom to live according to their needs. It is really an ironical fact.

Now a days, the knee problem is very common among aged people that creates hindrance in using Indian toilet seat because of their problem to sit down on the feet and folding their knees. It has been seen that the total number of accidents in home take place with the highest occurrences in the toilets, bathrooms, kitchens and stairs. In spite of that elderly have so far been treated as a problem group and issues about the elderly have seldom been listed to priority in planning. There are many factors which should be considered especially of third age people, like comfort, softness and support which they need in this age while using their toilets. Many of them are unable to sit on the Indian toilets due to their knee-joint problem and they are not able to stand after using the toilet without support but these factors are not considered by the designers.

The efforts of Dr. Nirmal Kaur, Dr. Promila Sharma, Gyanendra Sharma and Dr. R.C.S Mehta in compilation of this book by highlighting the features related to designing aspect for old age people entitled "Designing of toilet chair for elderly people" are appreciable. The book will help the researchers, construction agencies, furniture designers as well as manufacturers involved in this type of work to consider the limitations associated with elderly people while designing any product for them. I would like the book to be read by all those interested in improving the quality of life of elderly people.

(B.S. Bisht)

PREFACE

The increase in the number of old people is presumably also increasing the number of old people with disabilities. Disability in old age should be a focus of all public health policy and programmes and cover the whole continuum of services aimed at both preventing disability and restoring functional capacity through rehabilitative measures.

Ageing emphasizes problems in the environment and there is thus the need to create the right environment. The right environment in this context is when the older persons feel fit with their environments and in which they feel most comfortable

Changes in body composition are well known to occur with old age, but there is limited knowledge of the normal values rate of changes in anthropometric indices of body composition or the distribution of these changes in the elderly. The anthropometric standards derived from adult population may not be appropriate for the elderly because of body composition changes occurring during ageing. Specific anthropometric reference data for the elderly are necessary, before designing any kind of product or furniture or similar related equipment.

The major part of the research is based on the requirement of third age people regarding their toilet seat. Many companies are manufacturing toilet seats but there is no consideration for third age people having joint problem as a common illness in this age. There are so many factors which should be considered especially of third age people like comfort, softness and support which they need in this age while using their toilets. Most of the third age people are not able to move, thus they need folding type of toilet chair so that they can use it nearby their beds. Besides this there are families who are not able to construct European type toilets in their homes due to reasons like, space problem, financial problem, disliking by other family members or sometimes the family members do not bother about elders to take so much pain for them.

Just to solve this problem, there are locally made folding chairs available in the market but these are not up to the requirements of third age people, as these are not ergonomically designed and thus not suitable and also the comfortability is totally ignored. Their heights and widths are improper and there is no scope of support in any ways.

The design and psychological comfort and discomfort, thus may be estimated and need is to find out the ways to reduce it. The ill design of the toilet seat may produce degenerative changes in body. There is an urgent need to think about this important issue that is the requirement of third age people for functional, comfortable and safe toilet chair. Keeping, the above issue in mind, it is proposed that the toilet chair for third age people should be ergonomically sound, comfortable and should be constructed on certain guidelines by keeping in mind their special requirements.

Nirmal Kaur
Promila Sharma
Gyanendra Sharma
R.C.S. Mehta

Acknowledgements

"The satisfaction and euphoria that a company has in the successful completion of any task would be incomplete, without the mention of the people who made it possible and whose constant guidance and encouragement served as beacon of light and crowned the effort with success." Thus, we consider it as our privilege to express through the pages of this study a word of gratitude and respect to all those personalities who inspired us in the completion of this study.

We are grateful to Mr. V.K. Singh Associated Professor, Department of Mechanical Engineering, for making the design of toilet chair without that our research can't be progressed. We express our deep gratitude to all the old age people for their cooperation during the study.

We are highly thankful and obliged to the owner and workers of "Sanjay Orthoaids", Delhi for constructing the toilet chair, which will be great contribution for the third age people.

Nirmal Kaur
Promila Sharma
Gyanendra Sharma
R.C.S. Mehta

Contents

1

Introduction

"Old age is a new soul
With a self -transcendence=good"

—Sri Chinmoy

Alongwith death and taxation, getting old is one of those inevitabilities of life that cannot be avoided no matter how hard we try. Conventional wisdom suggests getting old is synonymous with declining mental and physical capacities. Getting past 70 is thus a time for retirement, reflection, nostalgia and living the quiet life. Not only this, but, despite the overwhelming inevitability of getting old, the process is something most people are uncomfortable and will often try their best to ignore or even reverse.

Yet spiritual teacher Sri Chinmoy has never been bound by conventional wisdom and approaching his 75 birthday still shows no sign of slowing down in his ongoing pursuit of self-transcendence. Newness is not a feature we associate with old age inspite of that Sri Chinmoy has created new records in weightlifting in the late ages of his life and there are so more exceptions in our country like Nobel Prize Winner Rabindranath Tagore began his painting career in the evening of his life. These examples, suggest that if we live in the heart then we can always have access to the fount of creativity and spontaneity no matter what our age.

It is well-known that the world's population is ageing, with more developed regions leading the process, because of increasing survival to older ages as well as smaller numbers of births (United Nations Population Division, 1998 revision). Consequently, the support of this ever expanding elderly population has become of increasing concern.

The Causes and Consequences of Major Health Problems

Approximately 20 per cent of people aged 70 years or older, and 50 per cent of people aged 85 and over, experienced difficulties in such basic activities of daily living as bathing, dressing, toileting, continence, feeding, and transferring from chair to bed. Difficulties are also common in carrying out instrumental activities of daily living such as using the phone, housekeeping, laundry, public transportation, taking medication and handling finances.

Several studies in the 1980s and 1990s sought to understand whether lowered mortality is accompanied by improvement or, on the contrary, deterioration of a population's health status (Freedman *et al.*, 2002 and Sulander *et al.*, 2003). Several studies found no covert changes in disability in the basic activities of daily living, such as bathing, dressing and managing toilet visits (Manton *et al.*, 1993, Alloire *et al.*, 1999 and Winblad *et al.*, 2001). The most severe types of disability is generally associated with long term care needs. On the other hand, some studies have demonstrated rather clear decreases in the ability to perform instrumental activities of daily living, such as household chores, shopping, and medication management (Manton *et al.*, 1993 *and* Crimmins *et al.*, 1997) and also certain basic physical task such as lifting, climbing stairs and walking (Allaire *et al.*, 1999).

In the study of Hirvensalo *et al.* (2000), it is again focused that ageing is associated with increasing difficulty in getting around on foot, both indoors and outdoors. One definition of impaired mobility is the inability to walk a specified distance, to climb stairs without assistance and to cross a street controlled by traffic lights in the time allotted. It is important that in the future, the needs of older people are taken into account in the design of services and the improvements in housing conditions and neighbourhood quality in order to facilitate the involvement of older adults in social and physical activities in their homes and neighbourhoods.

Now-a-days, the knee problem is very common among third of age people who creates hindrance in using Indian toilet seat because of their problem to sit down on the feet and folding their knees. Thus to avoid it seat height should be 20"—24" that will give better leverage in regaining a standing position and the fall and slip injuries can be checked which are associated with this knee problem. As it is quoted by Housdroff, 2001, that more than one-third of adults of ages 65 years and older fall each year.

It has been seen that the total number of accidents in the home takes place with the highest occurrences in the toilets, bathroom, kitchens and stairs.

Design consideration regarding Functional Areas of Home

The older one becomes the more likely he or she is to spend every moment at home. In the light of this knowledge, and taking into account the prolonging of human life, it is certain that the home environment issues will affect more elderly people and its implications nationwide. A home may generally be divided into number of discrete areas which include bedroom, kitchen, living room, bathroom and others. Each area has special functioning work task, and it is thus necessary to design functional areas which do not impede the profile of an ageing user. Most of us regard homes as a safe place, but accidents often occur in and around the home. The homes need to be designed to promote familiarity and orientation with the environment. An ergonomic approach and gerontology would improve the relationship between the ageing user and the home environment.

In this way age is a critical variable relevant to design considerations in human factors research and practices. This conclusion is found on three primary facts :

1. The number of older adults in developed and developing countries today is higher than ever and is increasing.
2. There are critical age related differences between younger and older adults which necessitate specific design considerations.
3. The proportion of older adults within the global workforce and of all users of systems and products is increasing steadily.

The elderly represent the fastest growing segment of populations throughout the world, with the distinctive feature of being a very heterogeneous group; a healthy 80 year old person is not comparable with a healthy 60 year old person, nor are healthy 80 year old persons necessarily comparable with each other in biological age. Indeed, the concept of functional or biological age should gain more consideration in the elderly. Within any single population, individual variation is increased due to variable rates of ageing from person to person and from physiological system to psychological system within the same individual. There are also special groups of elderly, such as the bed ridden or institutionalized.

Changes in body composition are well known to occur with old age, but there is limited knowledge of the normal values rate of changes in anthropometric indices of body composition or the distribution of these changes in the elderly. The anthropometric standards derived from adult population may not be appropriate for the elderly because of body composition changes occurring during ageing. Specific anthropometric reference data for the elderly are necessary, before designing any kind of product or furniture or similar related equipment.

Another consequence of the demographic shift is the need to give greater attention to the design of products and services for older consumers who now exercise the power of the so-called 'grey pond' (The Henley Centre and Design Age, 1997).

Counselling Services and Policies for Old People

There is an apparent need to develop counselling services for elderly people to assist them to better cope up with the multitude of questions related, for example, ways to counteracting the development of disabilities, managing chronic conditions, developing and maintaining healthy life styles, and being actively involved in social life.

The world is posing a new challenge for planners in dealing with an ageing society. This trend has implicated especially on planning requiring an accurate interpretation of an ageing society to accommodate their specific needs and wants.

The elderly have so far been treated as a residual or even a problem group and issues about the elderly have never been listed to priority in planning. The extent to which planners have seen older people as members of the community must be questioned in the light of the planning decisions which have placed less affluent and less mobile people at a disadvantage.

Disability in old age should be a focus of all public health policy and programmes and cover the whole continuum of services aimed at both preventing disability and restoring functional capacity through rehabilitative measures.

Rationale of the Study

The major part of the research is based on the requirement of third age people regarding their toilet seat. Many companies are manufacturing toilet seats but there is no consideration for third age people having joint problem as a common illness in this age.

There are so many factors which should be considered especially of third age people like comfort, softness and support which they need in this age while using their toilets. Many of them are unable to sit on the Indian toilets due to their joint problem and will not able to stand after using the toilet without support but these all factors are ignored by the designers. Most of the third age people are not able to move, thus they need folding type of toilet chair so that they can use it nearby their beds. Besides, this there are families who are not able to construct European type toilets in their homes due to reasons like, space problem, financial problem, disliking by other family members or sometimes the family members do not bother about elders to take so much pain for them.

Just to solve this problem, there are locally made folding chairs available in the market but these are not up to the requirements of third age people, as these are not ergonomically designed and thus not suitable and also the comfortability is totally ignored. Their heights and widths are improper and there is no scope of support in any ways.

Sitting is the most frequent body posture: we sit at

work, at school, in the car, on the bus, on the train, at home, in front of TV, to eat, to rest and in toilet and so on.

A seat should take the weight of your feet in order to lessen stress on your legs. It should provide some postural stability while you work, relax or in the toilet. You should be able to relax your muscles that aren't doing anything.

Generally, seats should allow your body to be comfortable and not restricted. The seat design is critical. The following facts should be taken into consideration for designing a seat:

- The seat height should not be so high so that your legs are left dangling! This would mean that there would be pressure on the soft tissues under your thighs. This pressure interferes with the return of blood from the lower limbs, which may cause tingling and numbness in the thighs which is due to pressure on blood vessels and nerves.
- The seat depth should allow clearance at the back of your knees in order to prevent pressure on the network of blood vessels and nerves.
- The seat back and angle should support the natural curves of your spine (in particular your lower back).
- The main weight bearing should be by the ischial tuberosities (the bony parts of your bottom) and the top half of the thighs.

In addition, a chair should enable the user to change posture at intervals, ensuring that different groups of muscles can be used for support, and that no particular group of muscles get tired. The consequences of poor seating are discomfort, fatigue and inefficiency in what one is doing.

The basic essential component is providing an appropriate environment for this particular population, whose distinguishing characteristics include impairment of basic functioning ability, lack of self-confidence, feeling of isolation and loneliness and lost identity. The design and psychological comfort and discomfort, thus may be estimated and need is to find out the ways to reduce it. The ill-design of the toilet seat may produce degenerative changes in body. Many studies were conducted on housing and elderly people but not many studies are undertaken specially in our country on ergonomic toilet chair and elderly people. There is an urgent need to conduct research on important issues that is the requirement of third-age people for functional, comfortable and safe toilet chair.

Keeping the above issue in mind, it is proposed that the toilet chair for third-age people should be ergonomically sound, comfortable and should be constructed on certain guidelines by keeping in mind the special requirements of third-age people. An ergonomically designed toilet chair should be friendly to human body and the dimensions of the chair should harmoniously match with each other. Hence, the present investigation was planned.

Objectives of the Study

1. To collect the anthropometric data of the people of third-age.
2. To study the clinical history of the people of third-age.
3. To study the occurrences of accidents among people of third-age.
4. Development of design through CAD for toilet chair.
5. Comparison between already developed toilet

chair available in the market and ergonomically designed toilet chair.

6. Validation of ergonomically designed toilet chair.

Hypothesis of the Study

1. The age and sex of third age people have no relationship with occurrences of accidents.
2. Anthropometric measurements of third age people have no relationship with the dimensions of toilet chair.
3. Anthropometrical measurements of third age men have no relationship with anthropometrical measurements of third age women.
4. Dimensions of simple toilet chair have no relationship with dimensions of ergonomically designed toilet chair.

Delimitation of the Study

1. The sample of the study was only the people of 60-70 years of age.
2. The sample was selected only from G.B. Pant University, Pantnagar and Rudrapur.

2

Review of Literature

The literature and reports of researches relevant to the present investigation are presented in this chapter under the following heads :

1. Anthropometry pertaining to furniture design
2. Functional limitation and disabilities found in third age
3. Health hazard/health problem among people of third age
4. Accidents like fall/slip among people of third age
5. Product designing for People of Third age
6. Conclusion

ANTHROPOMETRY PERTAINING TO FURNITURE DESIGN

Meaning of Anthropometry

The word "anthropometry" means measurement of the human body. It is derived from the Greek words anthropos (Man) and metron (measure). Anthropometric data are used in ergonomics to specify the physical dimensions of work spaces equipment, furniture and clothing so as to "fit the task to the man" (Grandjean, 1980) and to ensure that physical mismatches between the dimensions of equipment of products of the corresponding user dimensions are avoided.

Anthropometry is the subject, which deals with the measurements of the human external body dimensions in static and dynamic conditions (Chakrabarti, D. 1997). These include measurements of body parts, their strength, speed and their ranges of motion. External human body dimensional measurements are taken when a man is placed in a rigid and static position, whereas dynamic anthropometry is required more than static body measurements in the field of design. The human body is not rigid but rather, always dynamic.

Concern of Anthropometry for Sex Differences in Designing Products

Jeong and Park (1990) reported studies on sex differences in anthropometry for furniture design. The study was mainly concerned with anthropometric dimensions of persons and their interrelationship for furniture design. The study investigated sex differences in interrelationship between body dimensions to provide suitable sizes of furniture for males and females. The result showed that stature had high relationship to body dimensions for furniture design and there were significant sex differences in between stature and the body dimensions.

Evans *et al.* (1988), Kayis (1991) stated that anthropometric information for chair design is mainly concerned with providing data on stature of the people for whom the seats are designed.

The human geometry and the geometry of the product must be suitably fitted together (Kreifeldt, 1991) for safety as well as for functional reasons, hence, appropriate anthropometric and biomechanical applications are necessary for a "human compatible" designed product.

As a rule of thumb, if we take the smallest female of the tallest male in a population, the male will be 30 to 40 per cent taller, 100 per cent heavier and 500 per cent stronger (Grieve and Pheasant, 1992). Clearly, the natural variation of human populations has implications for the way almost all products and devices are designed. Some obvious examples are cloths, furniture and automobiles.

The design of furniture should be based on anthropometric characteristics and their relationship. It is also observed that there are sex differences in various anthropometric dimensions (Mandal, 1982; Jeong and Park, 1990 and Maudi, 1997). In order to provide suitable furniture to various sizes of persons the wider application of anthropometric characteristics with the sex difference is to be considered which will help to improve the seating and postural comfort of users.

Concern of Anthropometry for Geographic and Ethnic Differences

Different elderly populations show large geographic and ethnic variation in height, weight and Body Mass Index, much of which reflects differences in lifestyle and environment over the life course, genetic differences and to an uncertain extent, differences in health status (WHO, 1995). Currently available anthropometry data rarely include very old persons. NHANES II, the most comprehensive data set for anthropometry, does not include people older than 74 years. The committee of WHO encourages to collect anthropometric data on adults aged 60 years and above through anthropometric surveys conducted at regular intervals, coupled with the monitoring of the health and functional status of this segment of the population special attention should be paid to selection criteria in choosing population based samples, taking into consideration the heterogeneity of the elderly and the high prevalence of chronic conditions that may affect nutritional status. Several years hence, a consultation should be organized by WHO to review the current recommendations in light of available new data.

Concern of Anthropometry for Disabled

It is generally acknowledged that disabled populations show greater variation with regard to anthropometric data than able bodied populations. Different body measures of the normal sample was found broader than the range given in Pheasant (2001) for the ageing population.

Workplace furniture design and user anthropometry have become an important consideration in designing ergonomically appropriate furniture (Van Wely, 1970 and Harris *et al.,* 2005).

FUNCTIONAL LIMITATION AND DISABILITIES FOUND IN THIRD AGE

Meaning of Functional Limitation

As most studies have focused on the independent effects of various impairments, little information is available on risk for disability that results from the combined effects of multiple impairments, or as they are termed here, co-impairments. Functional limitations may be better explained by the accumulation of deficits across multiple domains rather than by any single specific impairments (Duncan *et al.*, 1993).

Types of Functional Limitations

The interaction between mobility limitation and environmental components adds further complexity to the disablement process (Shumway Cook *et al.,* 2003). People with mobility limitation often try to avoid travelling alone, climbing stairs, going to crowded places or using escalators. Walking limitation may also result from arthritis when pain predisposes people to decreased muscle strength and poor postural control.

A recent analysis showed that having high levels of fatalisms, low levels of self-confidence, and low levels of intellectual resources predicted greater difficulty in carrying out important life activities (Caplan *et al.*, 2003).

The prevalence of clinically important depressive symptoms among community dwelling older adults ranged from approximately 8 to 16 per cent (Blazer, 2003). For example, in the UK depression affects 10-15 per cent of people over 65 living at home (Anderson, 2002). Even higher figures have been reported (Heikkinen *et al.*, 2002). Depressive symptoms are more frequent among the oldest old, but the higher frequency is explained by factors associated with ageing, such as a higher proportion of women, more physical disability, co-morbidity, more cognitive impairment, and lower economic status. Many underlying causes of depressive mood have a social, economic or environmental background (e.g. emotional and social loneliness, poverty, poor living conditions and negative life events including losses of various kinds). It seems that depressive symptoms and disorders can result from functional limitations as well as cause them (Wells *et al.*, 1989; Penninx *et al.,* 1998 and Heikkinen *et al.,* 2004). When depressive symptoms are associated with sufficient disruption in everyday function, they tend to have a demonstrable public health impact (Johnson *et al.,* 1992 and Judd *et al.,* 1996). Despite a favourable response to treatment, late life depression remains largely undetected and untreated (Anderson, 2002 and Blazer, 2003).

In cross-sectional studies, physically inactive people often obtain poorer results in physiological and sensory motor tests, such as tests on aerobic capacity, muscle strength, postural balance, psychomotor speed and bone mineral content. Physically active people on the other hand maintain a higher level of physiological function, such as muscle strength, than their sedentary counterparts (Rantanen *et al.*, 1997). Exercises have been found to maintain good mobility and independence into old age (La Croix *et al.*, 1993). Among people surviving until very old age, even the period of disability prior to death has been found to be shorter among those, who were physically active earlier in life (Leveille *et al.*, 1999). Several studies have shown that emotional factors as well as cognitive factors are associated with functional limitations and disability among older adults *(e.g.* Laukkanen *et al.*, 1993 and Smits *et al.*, 1997). It has been suggested that psychological resources can strongly influence the relationship between personal capability and environmental demand by either increasing personal capability, reducing the environmental demand or both (Femia *et al.*, 1997).

Relationship between Functional Limitation and Dependence

Tinetti *et al.* (1995) found that when the number of impairments (lower and upper extremely, sensory and affective impairments) increased from zero to one, two and three or more, the proportion of participants experiencing functional dependence doubled from 7 to 14 per cent to 28 to 60 per cent, respectively. The impact of co-impairments of functional limitation and disability may be greater than the sum of those impairments singly. Cross-sectionally, the odds of severe walking limitation was ten times greater among those who had only one or the other (Rantanen *et al.*, 1999b). Among those who did not have severe walking limitation at baseline, during the three succeeding years the relative risk of onset of severe walking limitation adjusted for age, height, weight and race was more than five times greater in the group with the poorest balance and strength compared to the group with the best balance and strength. Among those who had good balance, strength was not associated with the risk for developing walking limitation, whereas among those with balance impairments good muscle strength decreased the risk for future walking limitation.

Functional Capacity v/s Life Expectancy

Almost all the indicators of physical and cognitive functioning seem to be related to life expectancy, showing that older individual with lowered functional capacity have a significantly shorter life expectancy compared to those with good functional capacity. In addition, the economic burden imposed on society by disability is great. It has been estimated that the use of social and health services increases in parallel with the increase in various disabilities (Rice *et al.*, 1992, Salive *et al.*, 1997 and Tsuji *et al.*, 1999).

Functional limitations were examined by Carmelli *et al.* (2000) using a lower extremity performance test. The heritability of the lower extremity test score was 57 per cent. Disability phenotype was studied in large Danish Twin Study using self-reported difficulty in three functional domains. The heritability of the scores varied between 34-47 per cent among women aged 80 years and over, while among younger women (75-79 years) the range was 15-43 per cent (Christensen *et al.*, 2000). These results suggest that genes play a role in the disablement process.

Relationship between Impairment, Functional Limitation and Disability

According to the disablement model, diseases are the main cause of impairments and functional decline which ultimately may lead to disability. People aged 70 years and over usually have 2-3 chronic conditions, which accounts for around two-third of total national health care expenditure (Lorig *et al.,* 1999 and Nilsson *et al.,* 2002). About 90 per cent of all 75 year olds have some clinical diagnosis. Interventions should, therefore, aim at the prevention and effective management of chronic illness (*e.g.* Grimley Evans, 2000). Also, sensory disorders and particularly vision and hearing impairments are important causes of problems in carrying out daily activities (*e.g.* West *et al.,* 2002).

Differences between Impairment, Functional Limitation and Disability

Impairment refers to the consequences of disease or injury to an organ system. The examples of impairment include muscle strength, vision, hearing or tactile sensitivity. Extensive information is available about various impairments as predictors of functional limitation and disability. The international classification of functioning, disability and health provider as exhaustive listing of these has been done (WHO, 2001). However, focusing on individual impairments has not proved to be sufficient in understanding disablement. A consideration of co-impairments appear to hold more promise.

Disability increases the risk of need for home help (Linden *et al.,* 1997 and Avlund *et al.,* 2001), hospitalization (Wolinsky *et al.,* 1994 and Avlund *et al.,* 2001), nursing-home admission (Sonn *et al.,* 1996 and Laukkanen *et al.,* 2000) and premature death (Sonn *et al.,* 1996 and Avlund *et al.,* 2002). Older people with difficulties in carrying out daily activities are in a danger of losing independence when placement in a nursing home becomes a realistic alternative (Laukkanen *et al.,* 2000). Such individual need help to be able to remain community dwelling. According to self reports by elderly people, disabilities feature among the most important determinants of diminution in quality of life.

Intervention Programmes for Prevention

There is an urgent need for effective intervention programmes for disability prevention, which has become an important public health concern (Ebrahim, 2002). The issues related to disability in old age cannot effectively be approached only through narrowly conceived interventions but rather by following the advice of Rudolf Virchow, "Mass problems require mass solutions". A comprehensive approach to disability prevention would focus on modifiable individual and environmental risk factors. There are however, both conflicting findings and gaps in our knowledge concerning the importance of various risk factors and the effectiveness of different kinds of interventions aimed at preventing and managing disability in old age. Further research work focusing on different aspects of disability in old age is therefore necessary for the formulation of successful policies and programmes (Heikkinen, 2003).

HEALTH HAZARD/HEALTH PROBLEM AMONG PEOPLE OF THIRD AGE

Physiological Problems

Barbaccia (1995) stated that the musculo-skeletal dimensions, mechanical performance, flexibility of joints, muscle strength, gait speed, bone densities are all important factors in the physiological system of changes which occur in these with ageing. Arthritis, heart disease, diabetes, difficulty with vision of hearing are common in older people. These problems with muscle of joints as hip fractures really contribute to a decrease in activities of daily living.

Health care professional have frequent personal contacts with older people, and are thus a significant resource for the promotion of regular physical activity. In particular, patients with ascertained risk factors or recently diagnosed diseases are candidates for behaviour change. However, only limited amount of evidence exists about the feasibility and effectiveness of exercise promotion in the context of health care organizations. Individualized physiotherapy may not be considered feasible owing to financial constraints. Sustainable forms of exercise include walking for fitness, home based exercise or supervised exercise groups. Some evidence exists that promoting more effective use of available resources, such as exercise classes provided by the community, is a cost-effective way to increase physical activity levels (Stewart *et al.*, 1997). A recommended strategy for health professionals in promoting physical activity among older people is to alter the key mediators of physical activity. Self-regulatory skills, social support, perceived benefits, motivational variables and self-efficacy have been identified as the main determinants of exercise adherence in previous studies (King *et al.*, 1998).

Advancement of Medical Technologies

The development of new medical technologies, which can successfully be used in the treatment of serious conditions in middle age (e.g. different coronary events), may in later life lead to an increased need for effective testing prevention and chronic disease management in order to prevent physical and mental decline in these patients. There is also evidence to suggest that a chronic disease self-management programme can improve health status while reducing hospitalization (Lorig *et al.*, 1999).

Even healthy elderly people may, through dietary modifications better maintain health, delay the onset of some diseases and prevent functional decline (Bales *et al.*, 2002). Nutrition education and counselling thus seem to be necessary in assisting older people to comprehend new information about healthy nutrition and apply it to their own situations (Sahyoun, 2002). Both obesity and underweight are common problems in older people.

There has been a continuous rise in the use of therapeutic drugs, especially in the industrialized world, which has also meant that an increasing number of drugs are being prescribed for elderly people. A study among community dwelling people aged 65-84 years showed that they used on an average 2.2 prescribed drugs per person and that one-fifth of the subjects had a greater number of drugs in simultaneous use (over 4 drugs) than is generally recommended (Laukkanen *et al.*, 1992). The increased use of psychotropic drugs (sedatives, neuroleptics, antidepressants) also increase unwanted effects such as confusion, accidental falls, memory decline and behavioural changes (Gurnack, 1997 and Armour *et al.*, 2002). The pharmacokinetics of drugs changes with age (*e.g.* Sadéna *et al.*, 2003), which given the high prevalence of co-morbidity in old age and the interactions of drugs, complicated the medication situation among this segment of the population.

Intervention Programmes

Specialized geriatric rehabilitation is complicated but effective when properly performed. Interdisciplinary teamwork, targeting of patients, comprehensive assessment and intensive and patient targeted rehabilitation seem to characterize the most effective programmes (Sletvold *et al.*, 1996).

There are many elderly patients with multi-pathology

functional limitations and related social problems. It is therefore important that interventions are also aimed at the effective management of chronic illness by providing both clients and service providers with adequate information, skills, incentives and resources. Evidence of the value (cost, well-being) of comprehensive assessment, optimized medical treatment, disease specific interventions, patient' education, physical exercise, nutritional advice and follow-up have been found in case of ischemic heart disease, heart failure, diabetes mellitus, stroke and in patients with hip fracture and complex geriatric conditions (Griffin, 1998; Anderson *et al.,* 2002; Joliffe *et al.,* 2002 and Lloyd-Williams *et al.,* 2002).

We also know that interventional strategies at the level of both the community and person can improve the quality of life of older people (Blumenhal *et al.,* 1999; Rybarczyk *et al.,* 1999 and Blazer, 2002). By focusing interventions on well known risk factors of depressiveness, adding support when needed, strengthening protective factors and improving coping skills, we can develop effective disability prevention strategies and at the same time promote mental well-being in late life.

A survey of the literature (Handoll *et al.,* 2003), including 30 scientific studies (9,496 patients) in randomized trails containing valid end points, has been able to show significant, reduction in mortality, postponement of nursing home placement, and some differences in outcome regarding function and costs.

ACCIDENTS LIKE FALL/SLIP AMONG PEOPLE OF THIRD AGE

Consequences of STF

Slip, Trip and Fall (STF) injuries present a tremendous human and economic burden in society (Leamon and

Murphy, 1995 and National Safety Council, 1995). The precise contribution of slipping and slipperiness to this burden are not completely understood but are assumed to be substantial.

A slip and fall related injury tends to be very rapid event lasting less than a second until, for instance and outstretched hand or the pelvis impacts with the ground. Hsiao and Robinovitch (1998) studied common protective movements associated with falls from standing height and found that a fall was more than twice as likely to occur after anterior translations of the feet (backward falls involving pelvic impact), when compared to posterior or lateral translations (forward falls). This backward fall mechanism appears to be similar to a common injury mechanism due to slipping after heel contact in walking (Strandberg, 1985; Gronqvist, 1999 and Leclercq, 1999).

Slip, Trip and Fall (STF) related morbidity and mortality are considerable in many developed countries. In the USA for example, slips and falls are the second largest source of unintentional injury mortality each year (Fingerhut *et al.*, 1998). Overall in 1997 in the USA falls were the leading external cause of medically attended, not fatal unintentional injuries with 11.3 million episodes reported at an age adjusted rate of 43.1 per 1000 persons (Warner *et al.*, 2000).

The multitude and complexity of extrinsic, intrinsic and mixed risk factors for slipping and falling, in fact, suggest that the injuries caused by slips are not the result of trivial incidents with simple prevention strategies (Strandberg, 1985 and Leclercq, 1999). On the contrary, even keeping track of the sequence of events following the initial perturbation (foot slide, loss of balance, or vice versa) is difficult. They are not fully understood due to their complexity and unexpected progression. However, several underlying perceptual, cognitive, psychological, biochemical, motor control and physical mechanisms seem to determine the onset and outcome of slipping accidents including the ways in which the human body reacts and adapts to a perturbation of balance and threat of falling (Chang *et al.*, 2001a,b,c; Gronqvist *et al.*, 2001 and Redfern *et al.*, 2001).

There are no undisputed methods to estimate slipping and falling risks, either in terms of the probability and severity of a loss of balance that can lead to a fall, or in terms of the probability and severity of a subsequent injury and disability (transient or permanent). As a result, the number and severity of fall and overexertion related injuries caused by slipping remain prodigious among the working population and among the general public, being a concern especially for the elderly population (Courtney *et al.*, 2001 and Gronqvist *et al.*, 2001).

Studies related to Accidents due to Slip

It is commonly assumed that slipperiness and slipping are major contributors to the STF injury burden. Across national boundaries, the registries that record injury data often differ substantially in their characteristics making international comparisons difficult *(Hagberg et al., 1997)*.

Foot slippage is the most frequent unforeseen event triggering falls on the same level and may also cause falls to a lower level (Anderson and Logerlof, 1983 and Courtney *et al.*, 2001).

Occupationally, Courtney and Webster (1999) recently reported that in the USA slips are associated with the most several disability, sudden onset occupational injuries including fractures. Leamon and Murphy (1995) estimated that annual, direct, per capita costs of occupational injuries due to slips and falls ranged from US$ 50 to US$ 400 per worker depending on the industry group considered.

Studies related to Accidents due to Fall

In the general population, it is estimated that one-third of community dwelling adults aged 65 years or older fall each year (Mc Elbinney *et al.,* 1998). Falls are the leading cause of death from injury in those aged 65 years and older and especially those over 75 years old (Sattin, 1992). Approximately, 9,500 deaths are attributed annually to falls of the elderly in the USA alone.

A fall to a lower level was in fact the major outcome in a survey on serious occupational accidents caused by slipping (Gronqvist and Roine, 1993).

Among older adults, falls are the leading cause of injury deaths (Murphy, 2000) and the most common cause of non-fatal injuries of hospital admissions for trauma.

Falls are common among the elderly and often lead to loss of independence or even death. According to a Yale University Study, the rate of falls among people aged 65 or older is about 30 per cent; among patients aged 80 or older, the rate is 50 per cent (Gill *et al.,* 2000), of the elderly who do fall, about 10 per cent sustain a serious injury such as a fracture, joint dislocation or severe head injury. Falls of fall related injuries are also associated with joint pain, loss of confidence, functional decline of placement in a nursing home. Most seniors want to "age in place" (Bayer and Harper, 2000). However, houses designed for families may be hazardous or downright unsafe for an ageing, frail person. Obstacles to renovation may include the stress of managing the project, finding competent of trustworthy contractions, and identifying financial resources (Steinfeld and Shea, 2001).

Falls precipitated by slipping are of major concern (Courtney *et al.,* 2001). Lloyd and Stevenson (1992) reported that slips and trips cause 67 to 32 per cent of falls sustained by the elderly and young, respectively.

Among older adults, falls are the leading cause of injury deaths (Murphy, 2000) and the most common cause of non-fatal injuries of hospital admissions for trauma.

Relationship of STF Injuries with Age

Kemmlert and Lundholen (2001) reported no differences by age in the exposure contributing to STF injury and suggested that occupational prevention strategies may not need to differ with age, while it is unclear exactly how the demographic trend of the ageing workforce (Robertson and Tracy, 1998 and Rantanen, 1999) will interact with all these factors, there can be a reasonable expectation that STF related occupational injuries in older working adults will be one of the expanding challenges in occupational safety in the twenty-first century.

Kenunlest and Lundholen (2001), reported that older were more likely to report greater proportion of STF injuries and to experience higher disability durations. The older people confronts challenges that can impact susceptibility to work related injury and disability including changes in body dimension, physiological and sensory capacities, and the ability to adaptively respond to physical stressor (Mital, 1994 and Laflamme and Menckel, 1995).

PRODUCT DESIGNING FOR PEOPLE OF THIRD AGE

Problems faced by Designers and Manufacturers

There are many age related changes in physiological sensory, perceptual, motor and cognitive abilities that may impact on how older people interact with their home related furniture. These changes include the following relevant declines in ability, decreased mobility, decreased strength, more rapid onset of fatigue, visual deterioration, reduced ability to process information slowed reaction time; and hearing loss (Smith *et al.*, 1993).

In commercial terms, it is crucial for designers and manufacturers to produce products, systems and services that are suitable for older and disabled users in mainstream design. Although, this is beginning to happen, progress is slow without the necessary legislation: the US currently has the strongest legislative climate; in the UK, although the Disability Discrimination Act (DDA) is leading to consideration of services and access to public buildings, there is currently no statutory requirement to design accessible products. A study by Ashworth *et al.* (1994) gives some indication of the extent to which older and disabled people are being 'designed out'. They report that 21 per cent of their sample of US 65-74 year olds and 55 per cent of 85 year olds, had at least some difficulty with activities of daily living (ADL's cooking, bathing and toileting, cleaning etc.) both at home and in the community. In addition, Clark *et al.* (1990) in their study of 60 older and disabled people aged between 55 and 93 found that 53 per cent were unable to shop, 45 per cent were unable to prepare meals and 28 per cent were unable to bath unaided.

As the users representatives in the product creation process, human factors specialists should concern themselves with wider issues than just usability. The integration of human factors throughout the product creation process is something which appears to be becoming organizations (Jordan *et al.*, 1995).

Traditional view of Product Designing

Traditionally, user centred design has tended to concentrate on 'usability'. Manufacturers increasingly see usability as an area where they can gain advantages over their competitors. This contrasts with the technical side of product development. Many manufacturing processes have now become so sophisticated that any advantages to be gained over competitors in terms of, say, price or product reliability are likely to be marginal (Jordan *et al.*, 1996).

Modern View of Product Designing

Richardson and Poulson (1996) believed that the application of methods, tools, techniques, and knowledge while dealing with questions of usability, would improve product and service quality. They also believed that it would also contribute to increased social integration, the quality of life, and independent living for ameliorating social consequences, and improving product and service market expectations. According to them, designing for the elderly and disabled people was designing for everyone.

The significance of knowing particular user characteristics is their possible relevance for the efficiency of subject sampling in user trialling. Generally, user trialling in a design context is not about the prediction of averages and dispersion in quantitative measurement results, but rather about the identification of different types of usages, especially in terms of difficulties in use (Kanis and Vermeesen, 1996).

As to everyday products, Green *et al.*, (1997), pose those in design the usefulness of human characteristics is largely limited to setting general boundary conditions, and then, there is very little in handbooks which seems to be used by designers in coming to grips with the actual activities of users in operating a particular design.

For setting general boundary conditions, the relevance is not questioned of acquiring, for instance, more anthropometric knowledge as a descriptor of new elements in a population which extend its variability, e.g. by ageing or by the arrival of ethnic groups. Another example is the importance of considering extreme intra-individual variation in force exertion, given a particular way of manipulation, in the adoption of adequate safety margins (Kayis, 1997).

Ergonomic data are fundamental to the design of safe and usable products (Norris and Wilson, 1997), and the benefits of using these data in the early stages of the design process are widely recognized. It was in response to this need for ergonomics data that the University of Nottingham, in association with the Consumer Affairs Directorate of the UK Department of Trade and Industry (DTI), recently produced a series of publications that bring together all available design related data into a compendium of easy to use design resources. The three publications on children, adults and older adults (Childata, 'Adultdata' and 'older Adultdata') (Norris and Wilson, 1995; Peebles and Norris, 1998 and Smith *et al.*, 2000) contain the most up to date anthropometric and physical strength data for countries around the world.

Observing people's interaction with products educates the designer on critical user behaviour than can serve as in toto design. An alternative to this is immersion in context which involves living with the users (Grossley, 2004). Behavioural patterns obtained using these techniques may not be quantifiable but give food cues about user needs in the natural setting. The end result is a richer design concept that encompasses user's conscious and unconscious aspirations about the product. This design concept however, can still be improved by using quantitative techniques to compare alternatives. It can also be tested for its potential to engender emotion.

Product design has always been in quandary as to the nature of product attributes that trigger users emotions. Emotions are compelling human experiences and designers can capitalize on this knowledge by conceptualize emotion engendering products that can sell well in the market. This entails knowing users' activities, thoughts, feelings, aspirations, goals, rituals and values and translating them into a product that elicits positive emotional response. Better designs are capable of provoking positive reactions from people such as sense of achievement, inspiration and joy (Giverchi and Velasquez, 2004).

Importance of Consulting Elderly

Coleman (1997) pointed out the importance of consulting elderly people during the design process, because they claimed that elderly people are surrounded by things which do not work well for them, or that they simply cannot find the things that they want. EUFRP (1997) is one of them, where years of experience, knowledge, and wisdom of the elderly were used to create new designs in the field of virtual communication and web use. Wolff *et al.,* (1997) conducted researches oriented towards the construction of housing for the elderly identified : senior housing, sheltered housing, nursing houses, community dwellings, and so on for old age people.

It is well known that the world's population is ageing, with more developed regions leading the process, because of increasing survival to older ages as well as smaller number of births (United Nations Population Division, 1998 revision). Consequently, the support of this ever-expanding elderly population has come of increasing concern. Another consequence of the demographic shift is the need to give greater attention to the design of products and services for older consumers who now exercise the power of the so-called 'grey pond' (The Henley Centre and Design Age, 1997).

A wide spectrum of professions is concerned with life

span design (universal design) for an ageing population, design, engineering, gerontology, ergonomics and architecture. All try to support ageing in place so that independence, freedom of choice and life-style are promoted. Designs considering the data related to both physical and psychological characteristics of people can improve the quality of life by promoting independence, as well as safety, usability and attractiveness of the residence (Demirbilek and Demirkan, 1998).

The elderly dealing with changed capacity, reduced ability and increased needs require the same accommodations and compensations in late life that they found in earlier years. Universal design is a concept that extends to a broad diversity of users who have to interact with the built environment (Sandhu, 2001; Scott *et al.*, 2001; Steinfeld and Danfort, 1993 and Story *et al.*, 1998).

Two surveys were conducted : one of UK practicing designers to investigate knowledge of inclusive design (Gyi *et al.*, 2000); and the other to understand the requirements and preferences of older and disabled people with regard to their interaction with everyday products, systems and services' (Oliver *et al.*, 2001). The Kansei Engineering framework was developed apparently to incorporate customers feeling into the design of the product by translating these feelings into the design elements (Matsubara and Nagamachi, 1997 and Nagamachi, 1995 & 2002).

The research of Aurelio (2000) has shown that many designers find it difficult to appreciate the full range of problems experienced by older and disabled people: HADRAIAN would facilitate this understanding. In addition, there are particular difficulties involving older and disabled people in design such as, high recruitment costs (e.g. time, transport costs); ethical consideration; and communication difficulties (Aurelio, 2000). It is not the intention to completely replace users' involvement in design but rather to complement existing methodologies. Many aspects of design and development do not lend themselves to be tested in a virtual environment anyway. Further details regarding the design tool itself can be found in Marshall *et al.,* (2002) and Porter *et al.,* (2004).

Conclusion

The extensive review of relevant literature made clear that a lot of studies have been done on common health problems faced by old age people, their functional limitations and disabilities and the accidents due to fall and slip etc. Furniture designing received considerable attention. Efforts are being made for designing different kinds of furniture for old age people. The design development for portable toilet chair was found an emerging field of research, as it was almost ignored by the designers. Though some kinds of toilet chairs are available in the market but that was not able to fulfil the requirements and limitations of this age group people. A few studies have been conducted in foreign countries. But it remains rather unexplored area in India. Therefore, a need was felt for development of ergonomically design for toilet chair. Hence, the present investigation was carried out.

3
Materials and Methods

The step by step procedure adopted to carry-out the investigation is presented in this chapter under various sub heads :

1. Research design
2. Conceptual framework
3. Selection of variables
4. Operational definitions
5. Development of the Tools for Data collection
6. Selection of the sample
7. Instruments used for Data collection
8. Data analysis

RESEARCH DESIGN

The main purpose of the present investigation was to study the problem among old age people regarding the use of Indian toilet and their requirement to replace this toilet with a portable toilet chair.

Hence, descriptive design was chosen to find out economic status of the family of the respondent, with the help of some questions like caste, education, occupation, income, possession and social participation of the family members through SES (Socio-Economic Status) scale developed by Singh and Saxena (1981).

Experimental design was planned to study anthropometric measurements for designing of toilet chair through CAD.

CONCEPTUAL FRAMEWORK

The conceptual framework to study the problems and accidents among old age people regarding the use of Indian toilet is shown in Figure 3.2(a).

2.1 The components of framework are:

1. Anthropometric measurements
2. Personal factors
3. Situational factors
4. Clinical occurrence of accidents
5. Design of toilet chair through CAD
6. Construction of toilet chair for user
7. Verification of toilet chair
8. Reduction in accidents

2.2 It is theorized that due to the ignorance among the family members as well as manufacturers regarding the limitations and problems of old age people related to their body measurements and hardness in the body due to age factor which leads to the occurrences of accidents either in the toilets or during toileting. Hence, they need an ergonomically designed toilet chair to avoid the cases of accidents.

SELECTION OF VARIABLES

There are two variables selected for the present research. The independent set of variables, acts through the dependent variables directly.

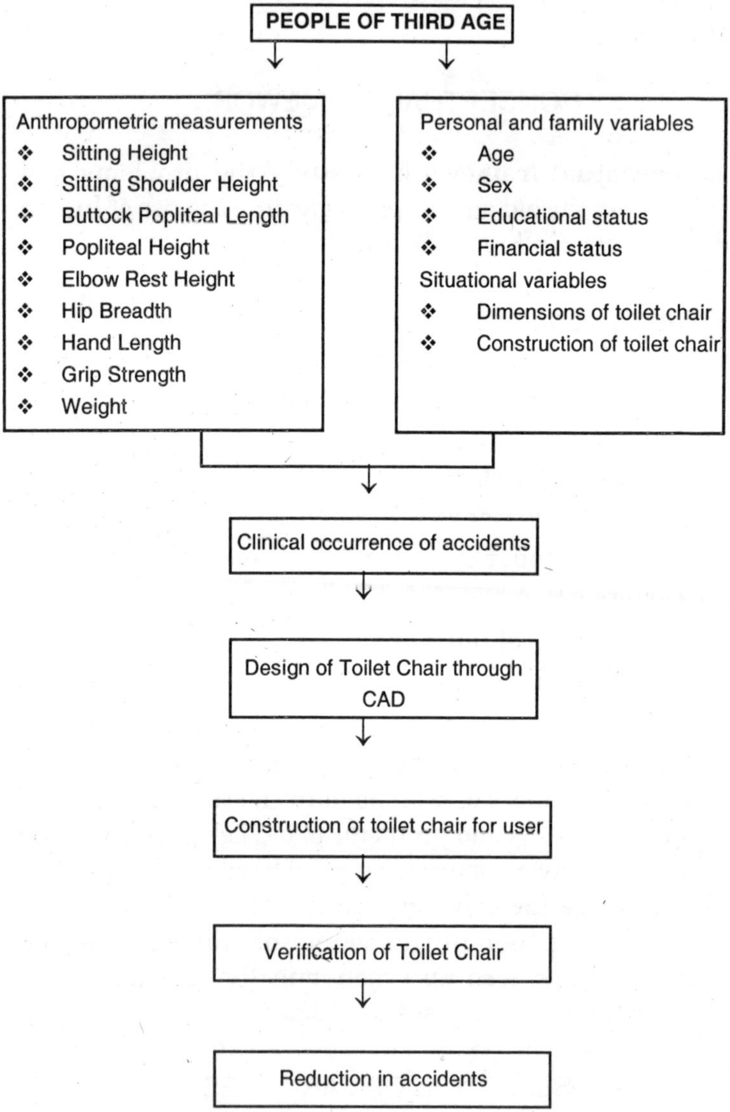

Fig. 3.1 : Conceptual framework to study the problems and accident among old age people regarding the use of Indian toilet.

Variables under consideration

Independent variables

A. Personal and family Variables

1. Age
2. Sex
3. Educational Status
4. Financial Status

B. Anthropometric measurements

1. Sitting measurements
2. Weight
3. Grip strength

C. Situational variables

1. Dimension of toilet chair
2. Construction of toilet chair

Dependent Variables

1. Ergonomical design of toilet chair
2. Reduction of accidents

Dependent Variables

In the present study, the extent to which ergonomical design of toilet chair causes reduction of accidents among old age people was considered as dependent variables. Each behaviour was considered separately as dependent or independent variables.

Independent Variables

For the present investigation, independent variables were classified in three categories, namely (1) Personal and Family Variables, (2) Anthropometric Measurements and (3) Situational Variables.

Personal and Family Variables

These include age, sex, educational and financial status of the old age people.

Age

Age was measured in terms of completed years of age at the time of study. Age of the third age people was assumed to have an impact on health hazard.

Sex

Sex was studied as number of male and female third age people considered for the study. There are differences in the body dimensions of male and female pelvic structures as well as in the body composition.

Educational status

Education is very important because it may affect the extent of knowledge of third age people having on the topic dealt with them, can also affect the health hazards.

Financial Status

It is very important because it can also affect the health of elderly people and it also predicts the facilities available to third age people.

Anthropometric Measurements

Anthropometrical measurements are often used in ergonomic design. Design development for comfortable working and seating involves anatomical, physiological and biomechanical considerations of the human body. Anthropometrical measurements have direct relation with the suitability and fitness of respondents to the furniture. The various sitting measurements taken for the study were :

Sitting Measurements

Sitting Height

This height was taken for deciding the height of backrest so that, if needed, the user can take the support for their head to keep upon that.

Sitting Shoulder Height

This height was taken for designing the backrest height of the toilet chair.

Buttock Popliteal Length

It was taken for the designing of seat depth. For this 5th percentile was taken so that the shortest person can also accommodate easily.

Popliteal Height

This height was taken for the designing of seat height. For the designing of adjustable seat height three dimensions were taken i.e. 5th percentile, 50th percentile and 95th percentile, so that all types of people means shortest, average as well as tallest can also sited easily on the chair.

Elbow Rest Height

For making the hand rest height appropriate, this height was taken and for this also 5th, 50th and 95th percentile was used.

Hip Breadth

This measurement was taken for the designing of seat width and for this 95th percentile was used so that broadest person can be fitted properly.

Hand Length

It was taken for the designing of hand rest and front handle rods circumference, so that the proper grip can be formed during sitting and standing.

Besides this weight and grip strength was also measured.

Grip Strength

It was measured for checking the power of the hands of people so that they can be able to carry the chair by their own.

Weight

Weight of third age people were measured for deciding the weight of toilet chair, so that it will be able to bear the weight of the heaviest person.

Situational Variables

Dimension of toilet chair

Discomfort and bad postures associated with poorly designed toilet chair may affect the health of third age people, therefore seat height, depth, width, backrest height etc. become an important aspect.

Construction of toilet chair

This is a very important variable which facilitates the comfort in using toilet chair that was ergonomically designed, like the dimensions as per the anthropometric measurements of third age people and all the safety parameters to avoid accidents.

OPERATIONAL DEFINITIONS

Certain terms were operationally defined for the measurement of variables of the present investigation which are described below.

Third age

Age is no more than chronological marker for the myriad changes that people undergo as the age. Third age starts from 60 years and end at 85 years.

Ergonomics

Ergonomics is a multidisciplinary subject, which can be applied to the study and design of human component of work systems. It provides a standardized approach to the analysis of work systems with emphasis on the interactions between human and machine. It deals with industrial medicine, engineering, psychology and system design and analysis.

Anthropometry

An empirical science defining measurement of person's size and form.

Sitting height

Vertical distance from the sitting surface to the top of the head, subject sits erect, looking straight ahead, with knees and ankles forming right angles (Fig. 3.2 A).

Sitting shoulder height

Vertical distance from the sitting surface to the highest point of the shoulder (Fig. 3.2 B).

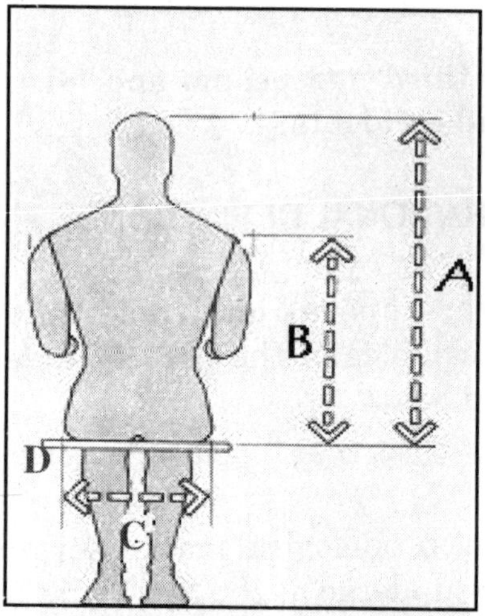

Fig. 3.2

Buttock-popliteal length

Subject sitting erect on a chair with knees flexed at 90° then it is the distance from the buttocks to the back of the knee (Fig. 3.3 E).

Popliteal height

Vertical distance from the floor to the underside of the thigh immediately behind the knee, subject sits erect, knees and ankles at right angles, bottom of thighs and back of knees barely touching the sitting surface (Fig. 3.3 F)

Elbow rest height

Vertical distance from the sitting surface to the bottom of the right elbow, subject sits erect, upper arm vertical at side, forearm at right angle to upper arm (Fig.3. 2 D).

Hip breadth (Sitting)

Maximum horizontal distance across the hips, subject sits

erect, knees and ankles supported at right angles, knees and heels together (Fig 3.2. C).

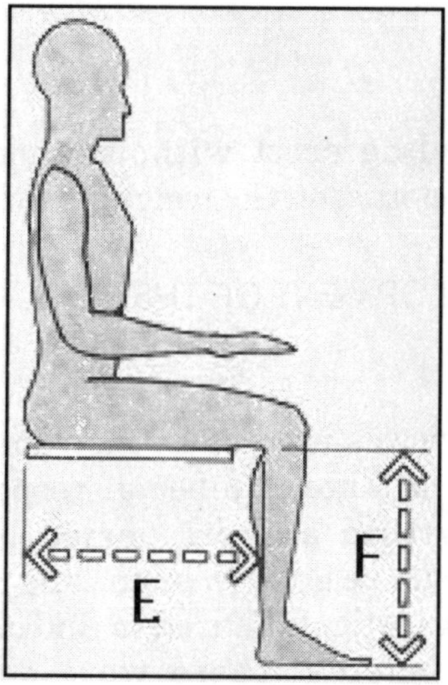

Fig. 3.3

Hand length

Distance from the base of the palm to the base of the middle finger. (Fig. 3.4)

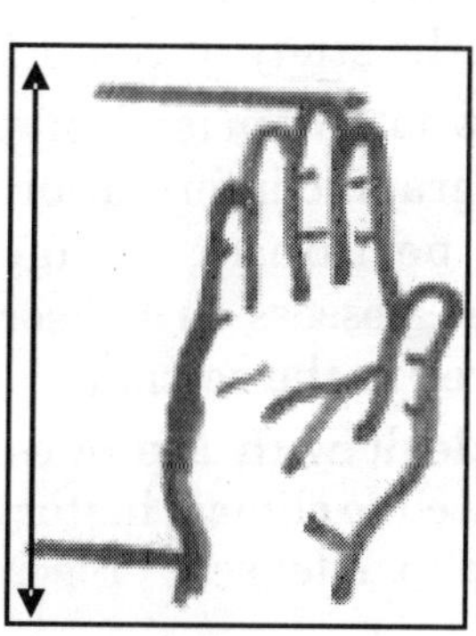

Fig. 3.4

Grip Strength

Person standing erect, keeping the hand horizontal to the body and holding the grip dynamometer with one hand at a time.

Weight

Person standing erect without shoes and with the minimum of cloths, on the weighing balance.

DEVELOPMENT OF THE TOOL FOR DATA COLLECTION

Questionnaire was used as an instrument for gathering data so that a large number of respondents could be contacted within a short period of time and the respondents can be given enough time to fill it up. Since, the present investigation tried to find out clinical history of the respondents, there were chances that the respondents would give their responses according to their recall capacity. Hence, the questionnaire was thought to be the most appropriate tool for the present study.

Description of the Tool

The Interview Schedule was constructed keeping in view the objectives of the study. It comprised of four sections :

Section I—was comprised of precoded questions regarding demographic information of the respondents. The questions pertaining to age, sex, occupation, education, income, possession and social participation and SES were involved in the section.

Section II—dealt with the questions regarding the information related to clinical history of the respondents, questions related to toilet seat present in their homes and all the interiors of the toilet and also the questions related to accidents if occur in the toilet.

Section III—dealt with the protocols for collecting the information regarding the anthropometric measurements of the respondents for designing the toilet chair.

Section IV—contained questions to validate the toilet chair in the form of checklist, regarding the performance of ergonomically designed toilet chair and satisfaction derived, were constructed in this section. The questions were given weightage and measured having 3 categories for the answers like highly satisfied having 3 weightage, satisfied having 2 weightage, and one weightage for not satisfied category. This checklist was given to the five experts of G. B. Pant University for content validity.

Development of Checklist for Toilet Chair Validation

An attempt has been made in the present investigation to develop a checklist to assess the practicability/lacunae of toilet chair.

Item Collection

The checklist comprised of statements (the items) relevant to the study. A thorough review of literature helped to frame the statements indicating the features of toilet chair, its adjustments, portability and others. The most important factor considered in collecting and framing the items was that it should be within the understanding of the respondents. Each item was thoroughly screened and edited so as to make it simple, clear and meaningful. The items were so framed that the respondents could say whether they were satisfied or not with the product made for them. In all, 15 statements were selected representing different aspects of toilet chair.

Content Validity of Toilet Chair's Checklist

The items framed were then, distributed to a panel of 5 judges who were experts from the G.B.Pant University, Pantnagar; members of the Advisory Committee actively working for the development of ergonomically designed products for different age groups. Objectives of the study and relevant operational definitions were supplied to them. The judges were requested to give their judgments on the correctness, clarity and relevance of each item. They were also requested to check each statement and indicate under which of the specified categories each statement exclusively fell. The responses of the judges were tabulated. The screening of the items was done on the bases of the following criteria.

All the items agreed upon by minimum 80 per cent of judges for clarity, relevance to the topic and positive/ negative framing were retained in the checklist. The items which were relevant but not clear, as reported by 80 per cent of judges, were rewarded and again given to three judges. If approved, then were included in the scale. Each item was considered to be exclusively falling under the category if reported so by 80 per cent of judges.

Out of the original set of 30 items, 25 items which fulfilled the above stated criteria were included in the toilet chair verification checklist for its validation.

Reliability of checklist for Toilet Chair Validation

The procedure followed in establishing the reliability of the checklist is reported below,

Scoring of responses on the checklist

The interview schedule was pre-tested on thirty old age people selected for pilot study, but was not included in the final sample. Their responses on each item of toilet chair validity were quantified by ascribing scores. The respondents were asked to indicate for each item whether it was correct, incorrect or they did not know. Thus, there were three categories. Correct responses were given a score of one and the "incorrect" as well as "I do not know" were given a score of zero. Incorrect answer was considered as good as having no satisfaction with the toilet chair design or material. There was thus, a possibility of a respondent scoring the minimum of zero if all answers were either wrong if they did not know and maximum of 25 if all the answers were correct as the checklist had 25 items.

Item analysis of checklist

The item analysis of checklist was done to eliminate inconsistency of the items in the checklist thereby increasing the reliability of the scale. The responses of the respondents were scored as described earlier. Thus, the weighted score for each item and also for each respondent was summed up. Correlation coefficient between the item scores and total scores of each respondent were computed by using the Pearson-Product-Moment formula. Item showing significant correlation values were selected for inclusion in the final checklist. Out of 25 items, 15 items were retained for the validation of toilet chair. Care was taken that the items selected finally covered or represented all the relevant aspects about toilet chair as listed earlier.

Reliability Coefficient of Toilet chair checklist

Split-half technique was applied to determine the reliability coefficient of the checklist. The whole checklist was divided into two halves employing odd-even method. Each of the two sets of items were treated as separate checklist. The respondents who scored high on odd items should score high on even items as well, if empirical errors have been kept to a minimum and the same applies in the case of low scores as well. The total score of each respondent in the odd and even categories were found out. The coefficient of correlation between odd and even score of thirty respondents was computed by the formula of Pearson-Product-Moment-Correlation-Coefficient. The value for the split half test was r = 0.77.

From the correlation of the half tests, the reliability coefficient of the whole test was estimated using Spearmen-Brown Prophecy formula which is

$$r = \frac{(\Sigma xy)}{\sqrt{(\Sigma x)^2\ (\Sigma y)^2}}$$

Following equation was used to understand reliability coefficient of the scales:

$$rrt = \frac{2rhh}{1 + rhh}$$

where,

r = Correlation co-efficient
rtt = Reliability of the total test estimated
rhh = Correlation between halves

SELECTION OF THE SAMPLE

The present investigation was carried out in Udham Singh Nagar district. As it was understood through the review of literature that old age people are being neglected by their family members and get suffered from accidents relate to the use of toilets. So there is a need to turn their attention towards their old parents and their immediate needs and other problems related to toilet seat. It was thought that people living in urban area, having good income as well as education might have turned their attention towards their old parents. Also, if alternatives are available, they would select the toilet chair which is good for their parents in terms of its function and material and dimensions. Such options are not available in near by market. Hence, it was thought appropriate to conduct research in this area.

Criteria for Selection of Sample

The respondents for the present investigation were men as well as women. Only those old age people were considered eligible to be included in the sample who—

1. Live in Udham Singh Nagar distict
2. Having joints problem
3. Have only Indian toilet in their home
4. Can't do construction due to any reason
5. Need European type portable toilet chair.

Collection of Data

The old age people living in Pantnagar and Rudrapur block of Udham Singh Nagar district were contacted for the collection of data by the investigator herself. Two hundred old age people out of which 50 per cent were male and 50 per cent were female were administered the interview through purposive sampling method. The period for data collection was from Semptmber 2007 tc February 2008.

INSTRUMENTS USED FOR DATA COLLECTION

Anthropometer

Anthropometric kit is a precision instrument made up of four interconnecting sections of tubular metals that are engraved in millimeter intervals. Current models are capable of measuring stature or other heights from the floor and seated surfaces as well as straight lengths and distances up to 200 cm when completely assembled. The heights were noted by using movable slide that contains an adjustable perpendicular blade, which is placed in alignment with tightened lightly on the desired measuring point.

The two upper sections of anthropometer may be used as a beam caliper as they were equipped with a millimeter scale on the slide opposite to the main scale that starts with zero mm at the top fixture. The beam caliper, which was capable of measuring distance up to 95 cm, is used for measuring whole body depths and breadth as well as many straight linear, distance between landmark points (Plate 3.1).

Plate 3.1: Anthropometric Kit

Weighing balance

It is used to measure the body weight in kilogram of an individual with minimum cloths and accessories on the body (Plate 3.2).

Plate 3.2: Weighing Balance

Grip dynamometer

It was used to measure the stress of grip muscles. It consists of a handle for hand grip connected with a spring to a pointer on the marked dial. The subject whose grip fatigue was to be measured was asked to pull the grip handle before and just after completing the activity and reading on the dial in kg is recorded (Plate 3.3).

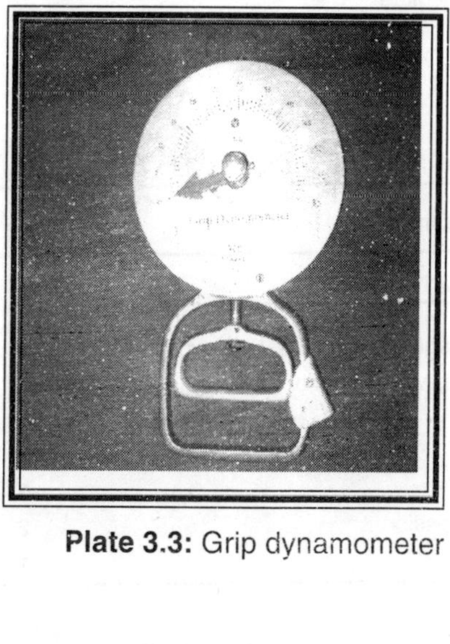

Plate 3.3: Grip dynamometer

DATA ANALYSIS

Categorization of the Variables for the purpose of Analysis

A.Socio-economic status

The scale of socio-economic status given by (Singh and Saxena, 1981) was used for this research and it includes education of the family members, their occupation, income, deposits and debts, type of house, material belonging, residential location etc.

The interview schedule was scored according to the weightage assigned to each item. The items and scoring are as follows:

(1) Education of family members

Categories	*Score*
University high degree	5
Postgraduate education	4
Graduate level education	3
Higher secondary, Intermediate	2
Middle school	2
Primary education	1
Literate	1
Illiterate	0

(2) Occupation of your family

Categories	*Score*
High administrative, Managerial or Professional work	7
Middle class Professional, Managerial or Technical work	6
Ordinary Profession	5
Skilled Profession	4
Service clerk	3
Middle class, Skilled Profession	2
Very ordinary work	1

(3) Income of the family

Categories	*Score*
Above Rs. 20,000	5
Between Rs. 15,000 - and 19,999	4
Between Rs. 10,000 – and 14,999	3
Between Rs. 5,000 – and 9,999	2
Between Rs. 1,000 – and 4,999	1
Rs. 999 or below	0

(4) Deposit and debt of the family

Categories	*Scores*	
	Deposit	*Debt*
Above Rs. 15001	5	0
Between Rs. 10,001 – and 15,000	4	1
Between Rs. 5,001 – and 10,000	3	2
Between Rs. 2,001 – and 5,000	2	3
Between Rs. 1,001 – and 2,000	1	4
Rs. 1,000 or below	0	5

(5) Sum of money collected at the time of emergency

Categories	*Score*
Above Rs. 15001	5
Between Rs. 10,000 – and 15,000	4
Between Rs. 5,001 – and 10,000	3
Between Rs. 2,001 – and 5,000	2
Between Rs. 1,001 – and 2,000	1
Rs. 1,000 or below	0

(6) Status of house

Categories	*Score*
Your own	3
A part of the house is on rent	2
Reside on rent	1
Reside on rent and have let-out your own house	0

(7) Type of house

Categories	*Score*
Big bungalow, big lawn, electricity and telephone	5
Small bungalow, small lawn and electricity	4
Pucca house electricity	3
Pucca house but quite small	2
Tile-shed, kachha floor	1
Kachha house	0

(8) Profession

Categories	*Score*
A	3
B	3
C	2
D	2
E	1
F	1
G	1
H	1

(9) Type of newspaper/magazine you purchase

Categories	*Score*
Daily	5
Weekly	4
Monthly	3
Quarterly	2
Sometime	1
Never	0

(10) Average expenditure on newspapers/magazine

Categories	*Score*
Rs. 100	1
Rs. 200	2
Rs. 300	3
Rs. 500	4
Above Rs. 500	5

(11) Shopping place where you go for

Categories	*Score*
Hatt	0
Ordinary shop	1
Mall	2
Branded showroom	3

(12) Social participation i.e. contact of elderly people

Categories	*Score*
Caste or community	3
Same profession	2
Political field	1
Neighbour and locality	0

(13) Opinion of your friends/other people about you and your family

Categories	*Score*
Most prosperous family	5
Prosperous family	4
Middle class	3
Average class	2
Low class	1

(14) Social status according to the members of family

Categories	*Score*
According to profession and work	3
On the basis of property and economic condition	2
According to caste	1

(15) Total SES

Class	*Score*
Lower SES	10-25
Lower Middle SES	26-38
Middle SES	39-52
Upper Middle SES	53-67
Upper SES	68 and above

B. Anthropometric measurement (in sitting position)

For anthropometric measurements 5th, 50th and 95th percentile were calculated and the following measurements were:

	Anthropometric measurement	*Calculated*
(1)	Sitting height	5th, 50th and 95th percentile
(2)	Sitting Shoulder Height	"
(3)	Buttock Popliteal Length	"
(4)	Popliteal Height	"
(5)	Elbow Rest Height	"
(6)	Hip Breadth	"
(7)	Hand Length	"
(8)	Grip Strength	"
(9)	Weight	Minimum, average and maximum

Statistical Analysis of Data

The data were coded and were analyzed employing descriptive as well as relational statistics.

Descriptive Statistics

The data were presented using frequencies, percentage, mean, percentiles and standard deviation and suitable tables and bar diagrams were also made for the classified data as per analysis of the following information:

1. Weight
2. Anthropometric measurement
3. Clinical history of joint problem
4. Type of material used in the toilet

Relational Statistics

The following null hypothesis were formulated and tested:

A. The sex of third age people has no relationship with occurrences of accidents.

B. Anthropometric measurements of third age people have no relationship with the dimensions of toilet chair.

C. Anthropometric measurements of third age men have no relationship with anthropometrical measurements of third age women.

D. Dimensions of simple toilet chair have no relationship with dimensions of ergonomically designed toilet chair.

Method of testing

- Chi-square test is used for testing the (A) & (B) hypothesis
- ANOVA is used for testing (C) hypothesis
- Fisher's T-test is used for testing (D) hypothesis

4
Results and Discussion

The facts and findings of the study emerged from statistical treatment of the collected data have been presented in the chapter. They are in concurrence with the said objective and are described under different sub-heads as follows:

1. Profile of the respondents.
2. Anthropometric measurements of respondents.
3. Clinical history of the respondents.
4. Assessment of occurrences of accidents among the respondents in the Indian toilet.
5. Assessment of interiors in the existing toilets of respondents.
6. Design development of toilet chair through CAD.
7. Comparison between the dimensions of ordinary and ergonomically designed toilet chair.
8. Validation of ergonomically designed toilet chair.
9. Testing of hypothesis.
10. Suggestions for modification in ergonomic toilet chair.

Profile of the Respondents

Significant result of selected respondents profile with respect to sex, age, educational qualification, occupation, monthly income, family status and socio-economic status of the families are described under this section.

Sex

Table 4.1 depicts that from total number of selected respondents from Pantnagar and Rudrapur, 50 per cent were male and 50 per cent were female.

Age

Body dimension changes as the age increases. So age was considered as an important variable (Grover, 2002). Table 4.1 indicates that out of 100 males, 64 males of third age were of 65-70 years and 36 males of third age were from 60-65 years of age whereas among females of third age, 62 females belong to 60-65 years of age and 38 were from 65-70 years of age.

Educational Qualification

Educational level affects the overall development of the members of the family. Table 4.1 shows that among males of third age, most of the respondents (i.e. 49 %) were found to have the education of higher secondary or intermediate and among females of third age most of them (i.e. 38 % were found to have middle school as the qualification.

Occupation of the Family

Out of the total sample, it was found that majority of the respondents belonged to middle class profession among both the categories (i.e. 60 and 63 %) males and females of third age, respectively. Whereas 15, 14 and 6 per cent males of third age belonged to ordinary profession, high administrative profession and skilled profession, respectively and among females of third age 16 per cent belonged to high administrative profession, 10 per cent were having ordinary profession and only 7 per cent were having skilled profession.

Income

Total income of the family affects economic stability as well as the purchasing power of the family. Table 4.1 revealed that most of the respondents, i.e. 53 per cent males and 62 per cent females of third age were having the monthly income between Rs. 10,000 to14,999, whereas there were families having the monthly income of Rs. 20,000 also, i.e. 27 per cent males and 18 per cent females of third age group families.

Family Status according to Others

Family status affects the report of a person in his society, because the economic factor is one of the major factors in present scenario, which affects the level of affordability of a person (Grover, 2002). Table 4.1 envisaged that out of total male families 47 per cent belonged to the middle class, 3 per cent to average class, 15 per cent were prosperous and only 4 per cent were found most prosperous in the eyes of others.

Whereas among females half of the families belonged to middle class and among remaining 50 per cent, 22 per cent were prosperous, 15 per cent were average class and 13 per cent were called most prosperous by the others.

Socio-Economic Status (SES)

Socio-economic status of the selected respondents was measured with the help of SES scale developed by Singh and Saxena (1981). The major component of scale was education of the family members, their occupation, income,

Table4.1 : Profile of the selected respondents

Personal Characteristics	*Male*	*Female*	*Total F=200*
1. **Sex**	100 (100)	100 (100)	200 (100)
2. **Age (years)**			
(a) 60–65	36 (36)	62 (62)	98 (49)
(b) 65–70	64 (64)	38 (38)	102 (51)
Mean	64.63	64.66	64.64
S.D.	9.59	3.22	7.13
3. **Education qualification**			
(a) University High Degree	8 (8)	2 (2)	10 (5)
(b) Post Graduation Education	13 (13)	3 (3)	16 (18)
(c) Graduate Level Education	11 (11)	12 (12)	23 (11.5)
(d) Intermediate or Diploma	49 (49)	7 (7)	56 (28)
(e) Middle School	14 (14)	38 (38)	52 (26)
(f) Primary Education	3 (3)	28 (28)	31 (15.5)
(g) Illiterate	2 (2)	10 (10)	12 (6)
4. **Occupation**			
High Administrative Profession	14 (14)	16 (16)	30 (15)
Middle Class Profession	60 (60)	63 (63)	123 (61.5)
Ordinary Profession	15 (15)	10 (10)	25 (12.5)
Skilled Profession	6 (6)	7 (7)	13 (6.5)
Service Clerk, Typist, P.A., Steno	1	1	2 (1)
Middle Class Skilled Profession	–	1	1 (0.5)
Very Ordinary Work	4	1	5 (2.5)
5. **Income per month**			
Above Rs. 20,000	27 (27)	18 (18)	45 (22.5)
Rs. 15,000–19,999	19 (19)	10 (10)	29 (14.5)
Rs. 10,000–14,999	53 (53)	62 (62)	115 (57.5)
Rs. 5,000–9,999	1 (1)	10 (10)	11 (5.5)
Mean	18385	15845.8	17115.4
6. **Family Status**			
Most Prosperous Family	4 (4)	13 (13)	17 (8.5)
Prosperous Family	15 (15)	22 (22)	37 (18.5)
Middle Class	47 (47)	50 (50)	97 (48.5)
Average Class	34 (34)	15 (15)	49 (24.5)
7. **Socio-economic status**			
Lower SES	2 (2)	3 (3)	5 (2.5)
Lower Middle SES	10 (10)	8 (8)	18 (9)
Middle SES	52 (52)	56 (56)	108 (54)
Upper Middle SES	28 (28)	27 (27)	55 (27.5)
Upper SES	8 (8)	6 (6)	14 (7)

Note : Figures in parenthesis represent percentage.

deposits and debts, type of house, material belongings and residential location, etc.

Table 4.1 envisaged that maximum per cent of the respondents of both the groups belonged to middle SES, i.e. 52 and 56 per cent from males and females of third age, respectively. It was also found that for both the groups upper middle SES, was also found in good percentage, (i.e. 28 %) and (27 %). It can also be seen in the table that there were few people of third age, which belonged to upper SES, as well as lower SES also.

Weight

As the age increases, the body weight increases to certain proportion then becomes constant (Grover, 2002). Body weight of the person is always measured with the standard weighing balance with minimum cloths and without shoes. For measurement of body weight direct method was used. Table 4.2 revealed that minimum weight among males of third age was 42 kg. and maximum was 96 kg, where respondents were found to have average body weight as 71 kg.

Table 4.2 : Weight of the selected respondents

Dimension	*Male*			*Female*		
	Min.	*Ave.*	*Max.*	*Min.*	*Ave.*	*Max.*
Weight (kg)	42	71	96	40	65	104

When it was measured for females of third age, then the results showed that the average women was about 65 kg whereas the minimum weight of females of third age were 40 and 104 kg, respectively.

Anthropometric Measurements of Respondents

Bone, muscle and other minute dimensions of anatomy

are mostly referred for medical and other necessities. But even while designing products for human use, the body dimension is of importance and the anthropometry has to be considered. The study of anthropometry is not just about body measurements, it is about how to employ body measurement information efficiently. While designing individual items or products, one has to take into consideration the dimensions of the product, their layout pattern in a given space, use etc, to match with anthropometry of the user. Anthropometry includes measurements of body parts, strength etc. Hence, in the present study some standard anthropometric measurements were taken by using direct method.

Sitting Height

Analysis of Table 4.3 exhibit that the mean sitting height, of males of third age was found as 80 cm and for females of third age it was 75.5 cm. When the 5th, 50th and 95th percentile was calculated for both males and females of third age, then it was found 72, 80.5 and 86 cm in case of males and about 71, 76 and 80 cm in case of females, respectively.

Sitting Shoulder Height

Sitting shoulder height for calculating the backrest height of the toilet chair and the mean for sitting shoulder height was found 56 and 54 cm for males and females of third age, respectively. Whereas, the 95th percentile was calculated as 61 cm for males, 59 cm for females. When the 95th percentile was calculated for both the groups, then it was found as 61 cm. (Table 4.3). For the purpose that both the groups can easily have the back support, the maximum height was taken 61 cm. Besides this, it was made adjustable, so that the respondents could adjust it as per their need as it was easy to move up and down.

Buttock Popliteal Length

All the values for Buttock Popliteal Length are shown in Table 4.3. The mean buttock popliteal length of the male and female of third age was measured as 45 and 44 cm, respectively and the 5th, 50th and 95th percentile were found as 40, 45, 53 cm, respectively for males, and 39.5, 44, and 49 cm, respectively for females of third age. As mentioned earlier, for the construction of toilet chair total score was calculated for all the measurements. For making, the proper depth for accommodation for both the respondents the 5th percentile value of both (male buttock popliteal length + female buttock popliteal length) was taken.

Popliteal Height

It is envisaged in Table 4.3 that the mean popliteal height for male of third age was 44 cm and for female of third age, it was 41 cm. Whereas the 5th, 50th and 95th percentile for male of third age was calculated as 40, 44 and 49 cm, respectively and the same for female was calculated as 36, 42 and 46 cm respectively.

Since it is not possible to make the toilet chair separately for males and females of third age, thus the whole data was calculated for the construction of toilet chair and for making the chair height appropriate for shortest, tallest as well as average height people. It was made adjustable at three heights as per the 5th, 50th and 95th percentile i.e. 39, 43 and 46 cm, respectively.

Elbow Rest Height

Table 4.3 display the mean elbow rest height of males and females of third age as 21 and 20 cm, respectively. Further

5th, 50th and 95th percentile were found as 16, 21 and 26 cm for males of third age and 15, 20 and 25 cm for females of third age, respectively

Hip Breadth

The mean hip breadth of the male as well as female group of third age was calculated as 38 cm. The 5th, 50th and 95th percentile were found as 34, 37 and 44 cm, respectively for males of third age and 30, 37.5 and 43.5 cm, respectively for females of third age. Whereas for toilet chair construction, the 95th percentile value of total measurement for hip breadth was taken (i.e. 44 cm) so that the broadest person could also be accommodated easily in the toilet chair.

Hand Length

Hand length was calculated for the designing of hand rest and front support rods circumference. By calculating the hand length of males and females, the 5th percentile value should be taken for designing, so that all the persons could have a tight grip over the hand rest rod as well as front support rod. It was found as 16 cm.

Grip Strength

It was measured for checking the power in the hands of third age people. It showed interesting results, which further showed the decline of power in ageing. In case of males of third age the mean value was 16 kg whereas in case of females of third age it was 12 kg. When the 5th, 50th and 95th percentile were calculated then it was found nil, 15 kg and 31 kg, respectively in case of males and nil, 9 kg and 31 kg in case of females of third age. The mean

Table 4.3 : Anthropometric measurements of Third Age People

Sitting Measurement (cm)	*Male*					*Female*					*Total*				
	M	SD	5th	50th	95th	M	SD	5th	50th	95th	M	SD	5th	50th	95th
Sitting Height*	80	4.9	72	80.5	8.6	75.5	3.05	71	76	80	78	4.69	71	77	86
Sitting Shoulder Height*	55	3.79	51	54	61	54	3.06	50	53	59	54	3.5	50	54	61
Buttock Popliteal Length	45	3.76	40	45	53	44	3.17	39.5	44	49	44	3.49	38.5	44.5	50
Popliteal Height*	44	2.65	40	44	49	41	4.88	36	42	46	42	4.1	39	43	46
Elbow Rest Height*	21	3.54	16	21	26	20	3.53	15	20	25	20	3.56	15	21	26
Hip Breadth*	38	4.06	34	37	44	38	4.43	30	37.5	43.5	38	4.24	33	37	44
Hand Length*	17	1.34	15.5	17.5	19	17	1.22	15.5	17	19	17	1.28	16	17	19
Grip Strength (kg)	16	10.4	0	15	31	12	9.68	0	9	31	13	10.2	0	11	31

* Significant Relationship.

value of male and female were calculated to see the amount of weight they could carry easily. It was calculated with the help of 5th, 50th and 95th percentile and found as 0, 11 and 31 kg, respectively. It depicts that they could carry maximum around 30 kg., because when a person carries something they can use both the hands simultaneously and also their half portion of the body (the upper half portion) gets involved.

Clinical History of the Respondents

Table 4.4 envisaged that 41 per cent of males and 62 per cent of females were having joint problem. It can also be seen from Table 4.4, that most of males as well as females developed this problem after 50 years of age whereas, it can also be seen that, some of the females (i.e. 8 %) reported this problem in their early life, i.e. at around 30 years of age and male also reported at around 40 years of age.

Due to joints problem the males (i.e. 39 %) felt pain during walking whereas 42 per cent females felt pain all the time. The reason may be that, they were having this problem for more than 2 years whereas; only 20 per cent males were having the pain for more than 2 years.

Efforts were made to discover the cause of problems and it was found that among the males the reason of their problem was fracture (i.e. 18 %), whereas, among females the common reason was slip (i.e. 33 %).

Efforts were made to find out the problems in using Indian Toilet and it was found that both the sexes were having pain in their joints and 28 and 14 per cent males and 32 and 21 per cent females had numbness in their feet and calf muscles, respectively. All the problematic males and females had problem in standing after sitting from Indian toilet.

Table 4. 4 : Clinical History of the respondents

	Criteria	*Male F=100*	*Female F=100*	*Total F=200*
1.	**Having joint problem**	41 (41)	62 (62)	103 (51.5)
2.	**Able to sit in squatting posture**	59 (59)	38 (38)	97 (48.5)
3.	**Problem in using Indian toilet**	41 (41)	62 (62)	103 (51.5)
4.	**Type of problem in using Indian toilet**			
	(a) Pain in knee joints	41 (41)	62 (62)	103 (51.5)
	(b) Numbness in feet	28 (28)	32 (32)6	0 (30)
	(c) Numbness in calf muscles	14 (14)	21 (21)	35 (18.5)
	(d) Standing after sitting	41 (41)	62 (62)	103 (51.5)
5.	**Time period of having problem**			
	(a) Last 6 months – 12 months	–	–	–
	(b) Last 12 months – 18 months	8 (8)	6 (6)	14 (7)
	(c) Last 18 months – 24 months	13 (13)	16 (16)	29 (14.5)
	(d) More than 24 months	20 (20)	40 (40)6	0 (30)
6.	**When feel pain**			
	(a) During standing	2 (2)	18 (18)2	0 (10)
	(b) During sitting	–	–	–
	(c) During walking	39 (39)	2 (2)	41 (20.5)
	(d) All the time	–	12 (12)4	2 (21)
7.	**How problem occur**			
	(a) After fall	4 (4)	5 (5)	9 (4.5)
	(b) After slip	6 (6)	33 (33)	39 (19.5)
	(c) After accident	13 (13)	4 (4)	17 (8.5)
	(d) After fracture	18 (18)	20 (20)3	8 (19)
8.	**Age when problem starts**			
	(a) Before 30 years	–	–	–
	(b) 30 +	–	8 (8)	8 (4)
	(c) 40 +	9 (9)	10 (10)	19 (9.5)
	(d) 50 +	32 (32)	44 (44)7	6 (38)
9.	**Treatment taking for cure**			
	(a) Medicine	8 (8)	12 (12)2	0 (10)
	(b) Physiotherapy	–	25 (25)	25 (12.5)
	(c) Exercise/Yoga	–	4 (4)	4 (2)
	(d) Avoidance of strain	33 (33)	–	33 (16.5)
10.	**Felt betterment**			
	(a) Better with treatment	4 (4)	2 (2)	6 (3)
	(b) Better after treatment	18 (18)	8 (8)	26 (13)
	(c) Better with physiotherapy	–	20 (20)2	0 (10)
	(d) Better after physiotherapy	–	5 (5)	5 (2.5)

Note : Figures in parenthesis represent percentage.

When respondents were asked about the treatment they were taking for cure of their problem then only 8 per cent males and 12 per cent females reported that they took medicines whereas, 33 per cent males expressed that they tried to avoid the strain and many females, 25 per cent go for physiotherapy and 4 per cent did exercise/yoga and 21 per cent females did nothing to cure their problem.

When people were asked about how they felt after treatment only 18 per cent males and 8 per cent females gave positive response for it and there were 4 per cent males and 2 per cent females who felt better with treatment. Among females there were 20 per cent females who felt better with physiotherapy and only 5 per cent were better after physiotherapy.

Assessment of Occurrences of Accidents among the Respondents in the Indian Toilet

Older persons, whose bones are often less dense and more brittle, are especially vulnerable to serious injuries from home accidents. Elderly people living independently without any assistance should consider all improved safety measures within their homes, as elderly people are more prone to serious injuries and long term disability if they fall.

Home injuries occur as a result of falls, slips and most of these accidents are considered preventable. Table 4.5 revealed age and sex-wise distribution of total number of accidents among the people of third age. It was found that all falls and accidents were due to the use of Indian toilet. Accidents were found more in case of females rather than males of third age, which results in injuries. It can be seen from Table 4.5 that these were increasing with the increase in the age.

Table 4.5 : Age and sex-wise occurrence of accidents

Criteria	*60-65 yrs*		*65-70 yrs*		*total*		*Chi-square values*
	M	*F*	*M*	*F*	*M*	*F*	
1. Fall due to standing after sitting	15 (15)	18 (18)	26 (26)	44 (44)	41 (20.5)	62 (31)	0.6466 NS
2. Fall due to no support	10 (10)	19 (19)	25 (25)	30 (30)	35 (17.5)	49 (24.5)	0.9404 NS
3. Fall due to slippery ness	4 (4)	7 (7)	8 (8)	11 (11)	12 (6)	18 (9)	0.0956 NS
Total	29	44	59	85	88	129	

Note : Figures in parenthesis represent percentage
M=Male, F=Female, NS=Non=Significant

Assessment of Interiors in the Existing Toilets of Respondents

It can be seen from Table 4.6 that out of total 200 males and females, 50 per cent were having Indian toilet in their homes and only 10 per cent males and females of third age were having European toilet and there were some males and females also (i.e. 40 %) who were having both types of toilet i.e. Indian as well as European in their home.

When they were asked about their preference of using toilet, then most of the males and females of third age (i.e. 60 %) preferred European type toilet. This indicated the cause of problem among third age people as they were not able to sit in squatting posture and this was the reason, they preferred to use European toilet. Besides this there were males and females of third age also who preferred Indian toilet (i.e. 30 %) as well as some of the males and females (i.e. 10 %) of third age reported that they were able to use both kinds of toilet.

It is envisaged from Table 4.6 that most of the males and females of third age (i.e. 92 %) were having raised platform of one step in their toilet, that is why, the third age people felt problem as they were suffering from joints problem. There were 63 per cent males and females of third age, whose toilet was in the level of the home. When the floor was checked then it was found that 87 per cent of the males and females of third age were having tiles on their toilet floor and 46 and 41 per cent males and females of third age were having marble and simple floor, respectively whereas, 26 per cent males and females of third age were having stone on the floor. About the door material it was seen that all the males and females of third age were having wooden doors in their toilet.

Table 4. 6 : Distribution of respondents as per the interiors present in their existing toilets

	Criteria	*Frequency*
1.	**Toilet present in the home**	
	Indian	100(50)
	European	20(10)
	Both	80(40)
2.	**Preference for the use of toilet**	
	Indian	60(30)
	European	120(60)
	Both	20(10)
3.	**Surface of toilet**	
	Raised Platform of one step	92(46)
	More than one stair	45(22.5)
	In the level	63(31.5)
4.	**Floor Material**	
	Simple	41(20.5)
	Marble	46(23)
	Stone	26(13)
	Tiles	87(43.5)

(Contd. .)

Table 4.6 : *(Contd...)*

5.	**Door Material**	
	Wood	200(100)
	Metal	–
	Glass	–
6.	**Window Material**	
	Wood	26(13)
	Metal/Glass	–
	Metal with glass	40(20)
7.	**Surface of wall**	
	Painted	46(23)
	White washed	94(47)
	Tiles	60(30)
8.	**Material of Toilet Seat**	
	Plastic	100(50)
	Stone	100(50)
9.	**Fixture in the Toilet**	
	Handle at one side of wall	–
	Handle at both side of wall	–
	Door handle	80(40)
	Tap	20(10)
	Wall	32(16)
10.	**Lighting in the Toilet**	
	Bulb	84(42)
	Fluorescent tube	24(12)
	CFL	–
	Combination of fluorescent tube + CFL	92(46)
11.	**Why European Toilet is not present in the home**	
	Financial problem	4(2)
	Family problem	10(5)
	Facility problem	–
	Space problem	50(25)
	Don't like	36(18)

Note : Figures in parentheses represent percentage.

Very few males and females of third age were having windows in their toilet. Twenty per cent were having windows of metal with glass and only 13 per cent were having wooden windows.

Among 23 per cent males and females of third age, surface of the wall was found painted in the home, whereas 47 and 30 per cent males and females of third age were having white washed and tiles, respectively on the surface of wall in their toilet.

Toilet seat was made up of plastic as well as stone in 50 per cent households.

In old age muscles loose strength compelling the aged people to seek support in sitting as well as in standing. But the findings suggest that none of the households surveyed, had any kind of support in their toilet. Forty per cent males and females of third age took support by holding door handle and 10 and 16 per cent reported that they usually hold the tap and side wall, respectively for support.

It was also found that lighting facilities were good enough inside the toilet as 42 per cent were having bulb, 12 per cent were having fluorescent tube and 46 per cent were having the combination of fluorescent tube and CFL. The reasons for not having European toilet in spite of its urgent need for them, was also investigated. It can be seen that 25 per cent males and females of third age were having space problem and 18 per cent males and females of third age were those, who did not like it whereas, only 5 and 2 per cent males and females of third age were those who had family problem and financial problem, respectively.

Design Development of Toilet Chair Through CAD

All the anthropometric dimensions of third age people were analyzed and on the basis of that, the final dimensions (shown in Table 4.7) were used for the design development of toilet chair for the people of third age.

Table 4.7: Dimensions for designing of toilet chair for the people of third age

Measurements	*5th Percentile*	*50th Percentile*	*95th Percentile*
Backrest Height (adjustable)	50 cm	54 cm	61 cm
Seat Height (adjustable)	39 cm	43 cm	46 cm
Seat Depth	–	44 cm	–
Seat Width	–	–	44 cm
Hand rest Height (adjustable)	63 cm	69 cm	74 cm
Rods Circumference	16 cm	–	–

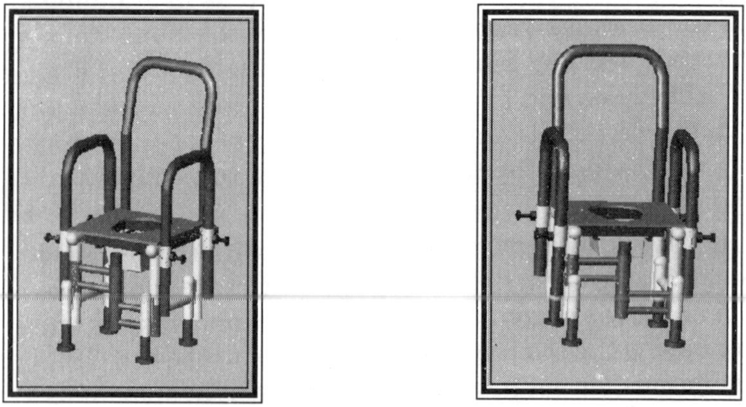

Plate 4.1 : Front View of toilet chair

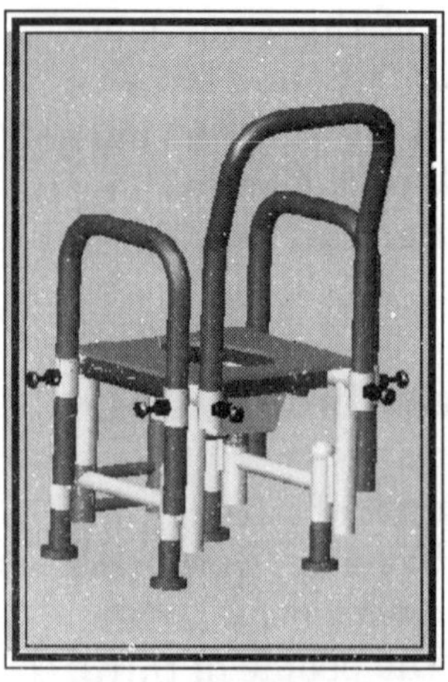

Plate 4.2 : Side view of toilet chair

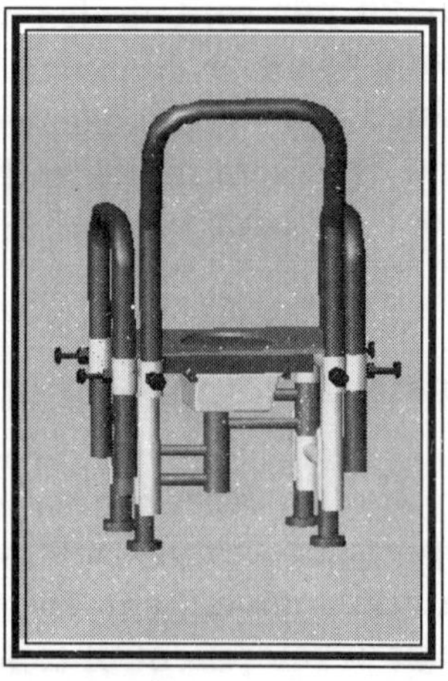

Plate 4.3 : Back view of toilet chair

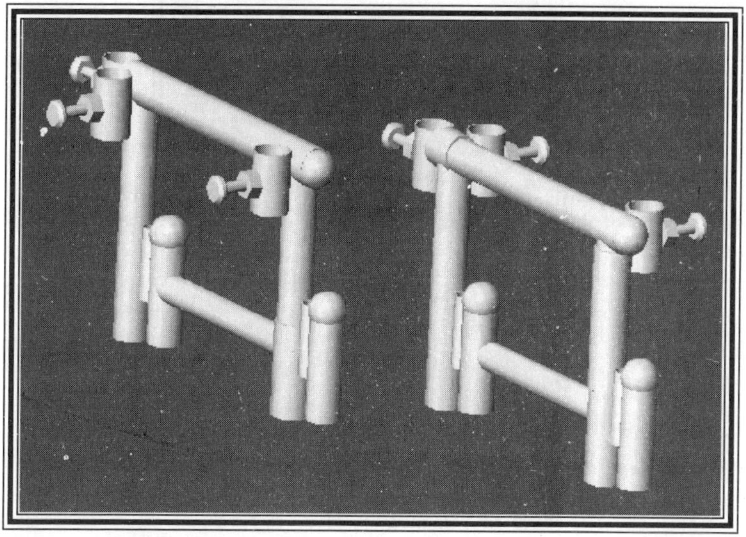

Plate 4.4 : Legs of toilet chair

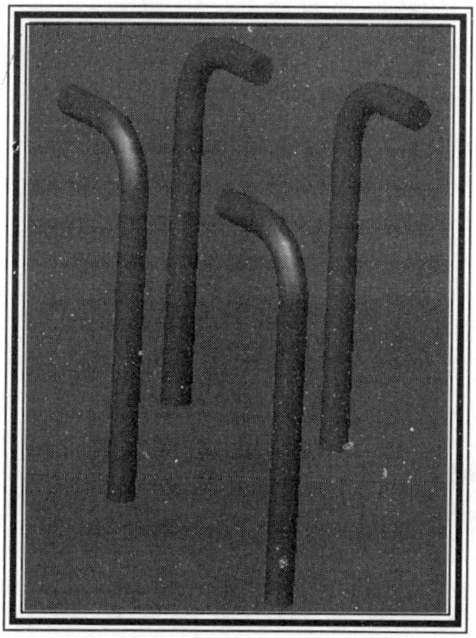

Plate 4.5 : Hand rest rods of toilet chair

Plate 4.6 : Assembled parts of toilet chair

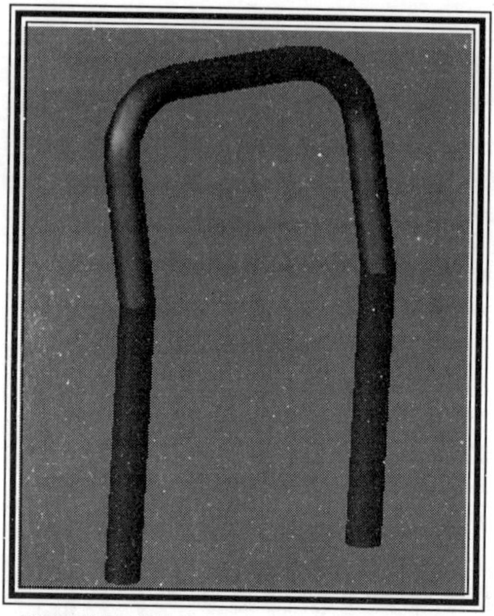

Plate 4.7 : Back rest rod of toilet chair

Plate 4.8 : Pan of toilet chair

Plate 4.9 : Seat plate of toilet chair

7. Comparison between the Dimensions of Ordinary and Ergonomically Designed Toilet Chair

When the dimensions of ordinary toilet chair, and ergonomically designed toilet chair were calculated for comparison then it was found (Table 4.8) that there was a measurable difference in seat height, seat width, backrest height and hand rest height of both types of toilet chairs. When it was also calculated statistically, than except for seat depth, a significant difference was found in all the dimensions of ordinary toilet chair and ergonomically designed toilet chair.

Table 4.8 : Comparison between the dimensions of ordinary and ergonomically designed toilet chair

Measurements (cm)	*Ordinary toilet chair*	*Ergonomic toilet chair*	*Fishers t-value*
Seat Height*	43	39	3.951
Seat Width*	40	44	3.360
Seat Depth	42	44.5	2.100
Back rest Height*	NP	61	35.218
Hand rest Height*	NP	26	30.0322
Weight of toilet chair*	2.5 kg	10 kg	25.0231

NP = Not Present * significant difference

As the ordinary toilet chair was very light in weight, the risk of disbalance was associated with it during sitting as well as during standing after its use. This was the most important reason that accidents might have occurred during its use among third age people.

Validation of Ergonomically Designed Toilet Chair

Extent of satisfaction derived by the people of third age was calculated on the controlled group of 20 people of third age. In order to get meaningful results, a checklist on satisfaction statements was constructed to assess the extent of satisfaction derived. Responses on each statement in the checklist were scored for highly satisfied (3), satisfied (2) and not satisfied (1). From the individual item score total and mean scores on each statement of third age people were computed.

The findings related to frequency distribution, total and mean scores on each statement of the scale are presented in Table 4.9.

With the height of the toilet chair, 70 per cent people of third age were highly satisfied whereas, only 30 per cent people were satisfied. Total score obtained was 54 and mean score was 2.7.

In case of weight of toilet chair, majority of the people of third age (i.e. 60 %) were satisfied, whereas, 25 and 15 per cent reported not satisfied and highly satisfied, respectively. In this case total score was found 38 and mean score was 1.9.

Table 4.9: Extent of satisfaction derived by third age people about ergonomically designed toilet chair

Description	*Highly satisfied = 3*	*Satisfied = 2*	*Not satisfied = 1*	*Total score*	*Mean score*
1. Height of chair	14 (70)	6 (30)	–	54	2.7
2. Weight of chair	3 (15)	12 (60)	5 (25)3	8	1.9
3. Width of chair	15 (75)	5 (25)	–	55	2.75
4. Depth of chair	15 (75)	5 (25)	–	55	2.75
5. Height of hand rest	20 (100)	–	–	60	3
6. Supporting handle	13 (65)	7 (35)	–	53	2.65
7. Operation of assembling	5 (25)	13 (65)	2 (10)4	3	2.15
8. Removal of pan	5 (25)	12 (60)	3 (15)4	2	2.1
9. Able to carry chair	3 (15)	9 (45)	8 (40)3	5	1.75
10. Material of toilet chair	15 (75)	5 (25)	–	55	2.75
11. Material of seat plate	4 (20)	11 (55)	5 (25)3	9	1.95
12. Material of back rest	5 (25)	15 (75)	–	45	2.25
13. Material of hand rest	14 (70)	6 (30)	–	54	2.7
14. Comfortableness	14 (70)	6 (30)	–	54	2.7
15. Cost of chair	5 (25)	10 (50)	5 (25)4	0	2.0

Note : Figures in parentheses represent percentage.

In case of width and depth of the toilet chair, there were 75 per cent people of third age, who were highly satisfied and only 25 per cent reported satisfied. The total score and mean score were found 55 and 2.75, respectively.

When the score for height of hand rest was calculated then it was found that 100 per cent people of third age were highly satisfied. Total score was calculated 60 with the mean score of 3.

There were 65 per cent people of third age who reported highly satisfied for supporting handle whereas, 35 per cent reported satisfied. Total and mean scores obtained were 53 and 2.65, respectively.

With the operation of assembling 65 per cent people of third age were satisfied, and 25 and 10 per cent were highly satisfied and not satisfied, respectively. Here, the total score was calculated as 43 and mean scores was 2.15.

Removal of pan was found satisfactory by 60 per cent people of third age and 25 per cent found it highly satisfactory whereas, only 15 per cent were there who were not satisfied at all. Total score and mean scores were 42 and 2.1, respectively.

When the people of third age were asked about whether they were able to carry the chair themselves, then there were only 15 per cent people who were highly satisfied and 45 per cent people were satisfied by it. There were 40 per cent people who were not able to carry the toilet chair by themselves. Here the total score was found 35 and mean score was 1.75.

Most of the people i.e. 75 per cent were highly satisfied by the material of the toilet chair and there were 25 per cent people who were only satisfied by it the total and mean score was obtained as 55 and 2.75.

Material of seat plate was found satisfactory by most of the people (i.e. 55 %) of third age and 20 per cent people reported that, they were highly satisfied and there were 25 per cent people, who were not satisfied at all. The total calculated value was 39 with the mean score of 1.95.

For material of back rest, 75 per cent people of third age reported satisfied and rest 25 per cent reported highly satisfied, with the total score of 45 and mean score 2.25.

In case of material of handle and comfortableness, 70 per cent people of third age reported highly satisfied and rest 30 per cent found it satisfied only. The total score and mean score was found 54 and 2.7, respectively.

There were 50 per cent people who were satisfied with the cost of the chair and 25 per cent people of third age were highly satisfied but there were 25 per cent people who were not satisfied with the cost at all. Total score was calculated as 40 and mean score was calculated 2.

The score earned by third age people was categorized in the ranges i.e. 0-15, 15, 30 and 30-45 scores and interpreted as low, moderate and high level of satisfaction, respectively.

Table 4.10: Distribution of third age people by their scores on extent of satisfaction derived

Score	*Range*	*Third age people*
0-15	Low	2 (10)
15-30	Moderate	6 (30)
30-45	High	12 (60)

Table 4.10 illustrates that 60 per cent people of third age were found to be in high level range, while 30 per cent in moderate range of satisfaction level. There were 10 per cent people who were in the range of low level of satisfaction.

Testing of Hypothesis

In order to test the hypothesis statistically the null hypothesis as presented below was formulated for each hypothesis.

The Age and Sex of Third Age People have no Relationship with Occurrence of Accidents

It was evident from the Chi-square calculated value that (in all the three categories listed above), there was no significant relation between the occurrence of accidents due to age as well as sex of third age people. It could be concluded that the occurrence of accidents and the age and sex of the third age people were independent factors.

Table 4.11 : Chi-square values for testing the relationship between occurrence of accidents with the age and sex of the third age people

Category	*Chi-square value*	*Conclusion*
Fall due to standing after sitting	0.6466	NS characters are independent
Fall due to no support	0.9404	NS characters are independent
Fall due to slipperyness	0.0956	NS characters are independent

NS = Not significant.

On the strength of three findings, the above null hypothesis H_0 was accepted and it could be inferred that the occurrence of accidents do not depend on the age and sex of third age people.

Anthropometrical Measurements of Third Age People have no Relationship with the Dimensions of Toilet Chair

It is evident from the chi-square calculated value that there was no significant relationship between selected anthropometric measurements of third age people and dimensions of ordinary toilet chair at 5 per cent level of significance. But the calculated value of chi-square was found significant in case of the dimensions of ergonomically designed toilet chair and selected anthropometric measurements of third age people.

Table 4.12 : Chi-square values for testing the relationship between anthropometrical measurements of third age people and dimensions, (a) ordinary toilet chair, (b) ergonomically designed toilet chair

Description	*Chi-square value*	*Conclusion*
1. Relationship between sitting shoulder height of third age people and		
(a) Ordinary toilet chair	0.000	NS characters are independent
(b) Ergonomically designed toilet chair	2.201*	Significant characters are dependent
2. Relationship between popliteal height of third age people and		
(a) Ordinary toilet chair	0.1025	NS characters are independent
(b) Ergonomically designed toilet chair	3.232*	Significant characters are dependent
3. Relationship between buttock popliteal length of third age people and		
(a) Ordinary toilet chair	0.0.763	NS characters are independent
(b) Ergonomically designed toilet chair	4.970	Significant characters are dependent
4. Relationship between hip breadth of third age people and		
(a) Ordinary toilet chair	0.909	NS characters are independent
(b) Ergonomically designed toilet chair	2.954*	Significant characters are dependent
5. Relationship between elbow rest height of third age people and		
(a) Ordinary toilet chair	0.153	NS characters are independent
(b) Ergonomically designed toilet chair	4.212*	Significant characters are dependent

NS = Not significant.

* Significant at 5% level

On the strength of these findings, the above null hypothesis $H_0 2$ was not rejected for ordinary toilet chair whereas for ergonomically designed toilet chair $H_0 2$ was rejected. Hence, it could be inferred that the dimensions of ordinary toilet chair were independent of anthropometric measurements of third age people whereas the dimensions of ergonomically designed toilet chair were dependent on anthropometric measurements of third age people.

Anthropometrical Measurements of Third Age Males have no Relationship with the Anthropometrical Measurements of the Third Age Females.

The anthropometrical measurements of males and females of third age were significant at both 1% and 5% level of significance. Only buttock popliteal length and grip strength of males and females of third age were found non-significant.

Table 4.13 : Correlation coefficients between anthropometric measurements of third age males and females

Correlation coefficient	*r value*	*Conclusion*
Sitting height of males and females of third age	0.435*	Significant at 5% level and 1% level
Sitting shoulder height of males and females of third age	0.312**	Significant at 5% level and 1% level
Popliteal height of males and females of third age	0.302**	Significant at 5% level and 1% level
Buttock popliteal length of males and females of third age	0.020	Not significant
Hip breadth of males and females of third age	0.990**	Significant at 5% level and 1% level
Elbow rest height of males and females of third age	0.385**	Significant at 5% level and 1% level
Hand length of males and females of third age	0.995**	Significant at 5% level and 1% level
Grip strength of males and females of third age	0.070	Not significant

** Signicant at 5% level and 1% level of Significance.

On the strength of these findings the above null hypothesis $H_0 3$ was rejected. Hence, it could be inferred that there was a significant positive correlation coefficient between the anthropometric measurement of third age males and females.

$H_0 4$: Dimensions of Ordinary Toilet Chair have no Relationship with the Dimensions of Ergonomically Designed Toilet Chair

It is evident from the *t*-value that there was no significant relationship between the dimensions of ordinary toilet chair and ergonomically designed toilet chair.

Table 4.14 : Fisher's *t*-value for testing significant difference between dimensions of ordinary toilet chair and ergonomically designed toilet chair

Description	*t- value*	*Conclusion*
Seat height of ordinary toilet chair and ergonomically designed toilet chair	3.951*	Significant
Seat width of ordinary toilet chair and ergonomically designed toilet chair	3.360*	Significant
Seat depth of ordinary toilet chair and ergonomically designed toilet chair	2.100	Not significant
Backrest height of ordinary toilet chair and ergonomically designed toilet chair	35.218*	Significant
Hand rest height of ordinary toilet chair and ergonomically designed toilet chair	30.022*	Significant

* Signicant at 5% level of Significance.

Hence, it could be inferred that the seat depth of ordinary toilet chair and ergonomically designed toilet chair showed equivalence whereas seat height, seat width, backrest height and hand rest height of both type of toilet chair showed significant difference. On the strength of these findings the above null hypothesis $H_0 4$ was accepted. It could be inferred as the dimensions of ordinary toilet chair were totally different from the dimensions of ergonomically designed toilet chair.

Suggestions for Modification in Ergonomic Toilet Chair

1. Alternative material should be found out, which can reduce the weight of toilet chair.
2. Pan depth should be increased.
3. Locking wheels can be used to make this toilet chair easily movable.

Conclusions

On the basis of findings of this investigation the following conclusions were drawn:

1. Among third age people, majority of third age belonged to 65-70 years of age and were having the education of higher secondary.
2. Among third age people majority of the people (males and females) belonged to middle class profession.
3. The third age people were having the monthly income between Rs. 10,000 Rs. 14,999 per month.
4. Regarding socio-economic status of third age people, majority of the males and females of third age belonged to middle class socio-economic status.
5. The anthropometric measurements of third age people were taken and on the basis of that, the total value was calculated for designing of the toilet chair. Correlation analysis indicates a significant relationship among the dimensions of males and females of third age.

6. For seat height adjustment, 5th percentile, 50th percentile and 95th percentile were used in the toilet chair so that all kinds of people i.e. shortest, average and tallest could be easily accommodated.
7. For seat width, 95th percentile and for seat depth, 5th percentile value was taken.
8. In hand rest and back rest the provision of adjustment was given. As the people of third age were having the problem of support, in the front, support handle was also given in the toilet chair.
9. Joint problem among the third age was found. Due to the problem in joints, majority of the people of this age felt pain all the time, and also during walking.
10. Third age people found problem while using Indian toilet, which caused pain in the joints among both the groups of third age. All the problematic males and females of third age were having problem in standing after sitting from Indian toilet.
11. Females of third age were more prone to accidental falls and slips than male. Most of the third age people went for physiotherapy to cure their problems. The findings of chi-square value showed the independence of occurrence of accidents on age and sex of third age people.
12. Regarding toilet conditions, it was found that lighting facilities was good enough. Wall condition was also found satisfactory. Floor which was made up of tiles was good enough but it was not found suitable for third age people as it was slippery for them.
13. None of the toilet had the support for third age people so that it could be used during standing after sitting.

14. The people of third age prefer to use European toilet.
15. Majority of the people of third age used the support of door handle for standing after sitting. The reason of not having European toilet was space problem.
16. The ordinary toilet chair was compared with the ergonomically designed toilet chair and the findings of Fisher's analysis showed a significant difference in all the dimensions except seat depth of both kinds of toilet chair.
17. The people of third age were found to be in high range of satisfaction in case of ergonomically designed toilet chair, as this toilet chair was designed by taking into consideration the actual requirement of the ultimate user. It was evident from the chi-square value of the third age people that there was no significant relation between the occurrence of accidents and the age and sex of the third age people.

Implications of the Study

The study brought out a number of important implications :

1. **For Designers :** This study will help those designers who are actually involved in designing of toilet chairs, about the importance of parameters which should be considered while designing for third age people.
2. **For Manufacturers :** The study will help those who are producing toilet chairs for third age people about kinds of features to be incorporated for these special need group.

3. **For Users :** This study will also help the users to make them aware about additional fixtures and support they want to incorporate in the toilet chair. The users can convey this information to the designers as well as manufacturers.

Scope for Future Research

1. The similar type of study can be conducted to discover alternative design and light weight material for production of toilet chair.
2. Study need to be carried out to develop some cushion material for seat plate which should be water proof in nature, so that the softness can be added in it.
3. Study can be done to find out what extra features can be added in these kinds of toilet chairs to enhance its comfortability.

Recommendations for Manufacturers

The word ageing is posing a new challenge for planners in dealing with an ageing society. This trend has implicated especially on planning, requiring an accurate interpretation of an ageing society to accommodate their specific needs and wants.

The elderly have so far been treated as a residual or even a problem group and issues about the elderly have never been listed to priority in planning. The extent to which planners have seen older people as members of the community must be questioned in the light of the planning decisions which have placed less affluent and less mobile people at a disadvantage.

Now-a-days, the knee problem is very common among third age people which creates hindrance in using Indian toilet seat because of their problem to sit down on the feet and folding their knees. Thus to avoid it, seat height should be 20-24" that will give better leverage in regaining a standing position and the fall and slip injuries can be checked which are associated with this knee problems. As it is coated by Housdroff (2001), that more than one third of adults' ages 65 years and older fall each year.

It was found from the study that there were so many families of middle income group who were not having European type toilet in their homes. So many reasons like financial problem and space problem were there due to which those families were not able to construct European toilets in their home.

To arrange something, the family members of old age people were making toilet chair for their parents in the home itself. They were making it by the use of ordinary chair available in their home. They simply cut the chair seat in between and put a small bucket in it.

Changes in body composition are well known to occur with old age, but there is limited knowledge of the normal values rate of changes in anthropometric indices of body composition or the distribution of these changes in the elderly. The anthropometric standards derived from adult population may not be appropriate for the elderly because of body composition changes occurring during ageing. Specific anthropometric reference data for the elderly are necessary, before designing any kind of product or furniture for elderly.

There are so many other factors also which should be considered especially of third age people like comfort, softness and support which they need in this age while using their toilets. Many of them are unable to sit on the Indian toilets due to their joint problem and well as not able to stand after using the toilet without support but these all factors are ignored by the designers. Most of the third age people are not able to move, thus they need folding type of toilet chair so that they can use it nearby their beds.

Just to solve this problem there are locally made folding chairs available in the market but these are not up to the requirements of third age people, as these are not ergonomically designed and thus not suitable and also the comfortability is totally ignored. Their heights and widths are improper and there is no scope of support in any ways.

Generally, seats should allow users' body to be comfortable and not restricted. The seat design is critical for this:

- The seat height should not be so high so that the user's legs are left dangling! This would mean that there would be pressure on the soft tissues under the thighs. This pressure interferes with the return of blood from the lower limbs, which may cause tingling and numbness in the thighs due to pressure on blood vessels and nerves.
- The seat depth should allow clearance at the back of the user's knees in order to prevent pressure on the network of blood vessels and nerves.
- The seat back and angle should support the natural curves of the user's spine (in particular your lower back).
- The main weight bearing should be taken by the ischial tuberosities (the bony parts of your bottom) and the top half of the thighs.

In addition, a chair should enable the user to change posture at intervals, ensuring that different groups of muscles can be used for support, and that no particular group of muscles gets tired. The consequences of poor seating are discomfort, fatigue and inefficiency in what the user is doing.

Keeping the above issue in mind, it is proposed that the toilet chair for third age people should be ergonomically sound, comfortable and should be constructed on certain guidelines by keeping in mind the special requirements of third age people. An ergonomically designed toilet chair should be friendly to human body and the dimensions of the chair should be harmoniously matched with each other. Hence, the present study entitled "Development of Ergonomically Designed Toilet Chair for the People of Third Age" was planned.

Design Criteria for Furniture Manufacturer

Furniture	*Design Criteria*	*Recommended Design Values (cm)*
Chair seat height	5th percentile popliteal height as well as 95th percentile popliteal height with adjustments at every 2 cm in between	39 cm,46 cm and 3 holes in between
Chair Seat width	Hip Breadth, 95th percentile	44 cm
Chair Seat depth	Buttock popliteal length,50th percentile	44.5 cm
Backrest height above the seat	Sitting shoulder height, 95th percentile and also adjustable	61 cm
Hand rest height	Sitting elbow rest height,5th percentile and also adjustable	63 cm
Supporting handles	Sitting elbow rest height,5th percentile and also adjustable	63 cm
Handles circumference	Palm length, 5th percentile	16 cm
Chair material	Iron with powder coating and all the edges are mounted with plastic cap	

Recommendations for Users

It is recommended that consumers should use ergonomically designed toilet chairs which are functionally sound and the material used in such chairs are eco-friendly.

These chairs have no ill effect on the health of users and no negative impact on the posture of users. These chairs are strong enough and stable thus, reduces risks of falls and slips.

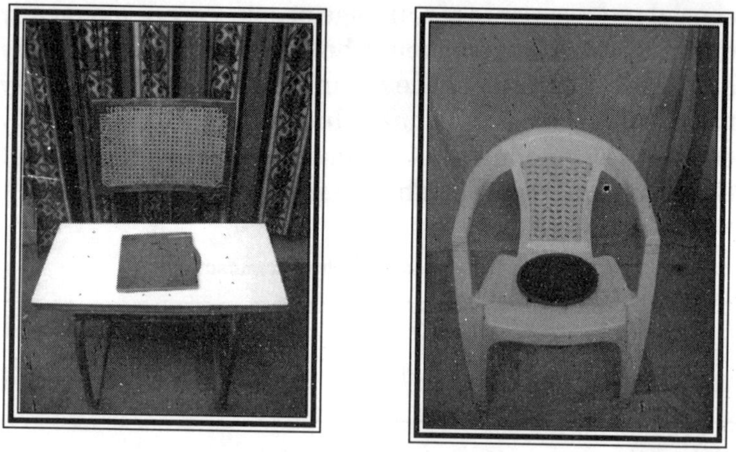

Plate 4.10 : Home made toilet chairs

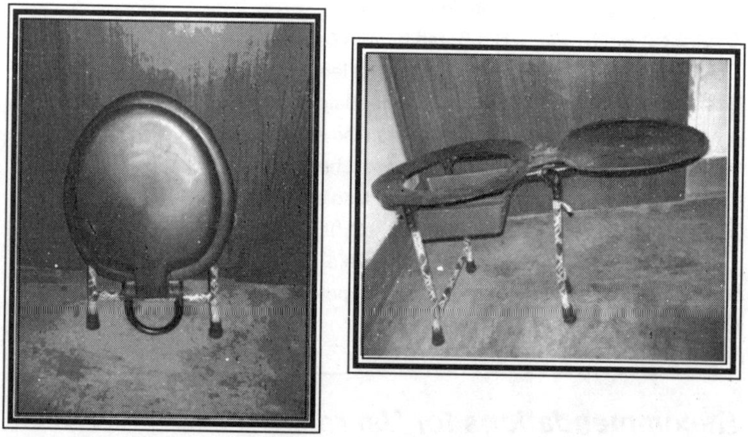

Plate 4.11 : Ordinary folding commode available in market

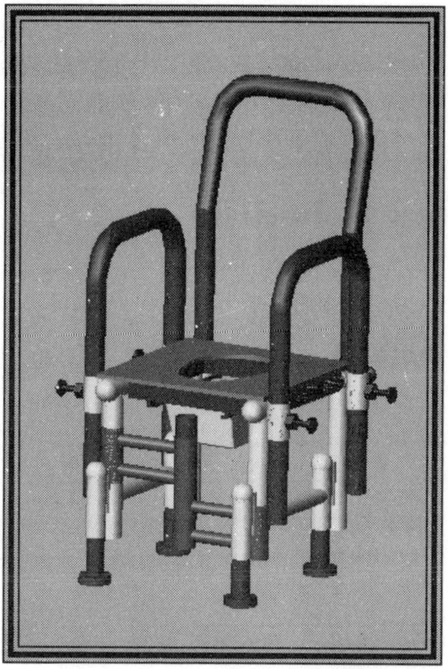

Plate 4. 12 : Design developed

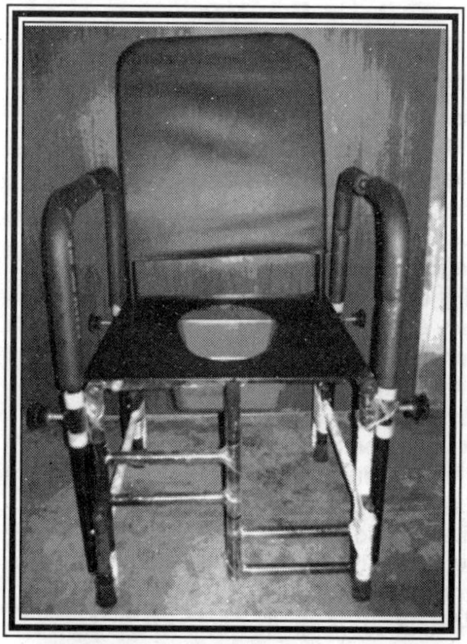

Plate 4.13 : Ergonomicall designed toilet chair

Appendix

Interview Schedule

PART-A

1. **Name of the respondent** ....
2. **Sex M**............................... **F**...
3. **Age**...
4. **Place**...
5. **Educational qualification**

Categories	*Respondent*	*Spouse*	*Son*	*Daughter*
A. University high degree e.g. Ph.D., Lit., M.S., M.D. or Similar professional degree	( )	( )	( )	( )
B. Post Graduate Education (M.A., M.Sc., M.Com., M.Ed.)	( )	( )	( )	( )
C. Graduate Level Education (B.A., B.Sc., B.Com., B.Ed., L.L.B etc) or any other equivalent degree after intermediate	( )	( )	( )	( )
D. Higher Secondary, Intermediate, any other professional certificates, or any other diploma after High School	( )	( )	( )	( )
E. Middle School (Class VIII passed) or equivalent training certificate	( )	( )	( )	( )
F. Primary Education (Class V passed)	( )	( )	( )	( )
G. Literate (did not pass classes but know little reading and writing)	( )	( )	( )	( )
H. Illiterate	( )	( )	( )	( )

6. Occupation

Categories	*Respondent*	*Spouse*	*Son*	*Daughter*
A. High Administrative, Managerial or professional work, such profession in which high education is required e.g. High Court, Officer, Proprietor of professional institutions, Advocate, Univ. Professor	( )	( )	( )	( )
B. Middle class professional, Managerial or Technical work e.g. Middle class Advocate, Shopkeeper, College Lecturer etc.	( )	( )	( )	( )
C. Ordinary professional, such profession which requires education up to B.A., e.g., Teacher of high school classes, Technician	( )	( )	( )	( )
D. Skilled profession, e.g., Electrician, carpenter etc.	( )	( )	( )	( )
E. Service clerk, Typist, P.A., Steno, Policeman etc.	( )	( )	( )	( )
F. Middle class skilled profession	( )	( )	( )	( )
G. Very ordinary work servant, labourer, peon, etc.	( )	( )	( )	( )

7. What is the Income of the Family?

A. Above Rs. 20,000 ( )
B. Between Rs. 15,000 and 19,999 ( )
C. Between Rs. 10,000 and 14,999 ( )
D. Between Rs. 5,000 and 9,999 ( )
E. Between Rs. 1,000 and 4,999 ( )
F. Rs. 999 or below ( )

8. What is the Deposit and Debt of Your Family?

	Deposit	*Debt*
A. Above Rs. 15,001	…………	…………
B. Between Rs. 10,001 and 15,000	…………	…………
C. Between Rs. 5,001 and 10,000	…………	…………
D. Between Rs. 2,001 and 5,000	…………	…………
E. Between Rs. 1,001 and 2,000	…………	…………
F. Rs. 1,000 or below	…………	…………

9. If Your Family requires an Urgent Some of Money how much Some of Money could be Collected?

A. Above Rs. 15,001 ( )
B. Between Rs. 10,001 and 15,000 ( )
C. Between Rs. 5,001 and 10,000 ( )
D. Between Rs. 2,001 and 5,000 ( )
E. Between Rs. 1,001 and 2,000 ( )
F. Rs. 1,000 or below ( )

10. The House in which you Reside Is:

A. Your own ( )
B. A part of the house is on rent ( )
C. Reside on rent ( )
D. Reside on rent and let out your own house ( )

11. How is Your House?

A. Big bungalow, big lawn, electricity and telephone ( )
B. Small bungalow, small lawn, electricity ( )
C. Pucca house, electricity ( )
D. Pucca house but quite small ( )
E. Tile-shed, Kachha floor ( )
F. Kachha house ( )

12. Tick (√) the Articles which you Posses at Home.

A. Car.................. Motor cycle or Scooter........... Bicycle................
B. Stereo................ Television............... Record player.................. CD player.......................
C. Refrigerator.................. AC.............PC...........Laptop.............
D. Sofa set.........................Good Chair Table..................... Wooden Chairs................ Dinning Chair...................
E. Ordinary Chair Table...............Wall Clock............Wrist Watch......
F. Gas burner............Electric Stove..........Dish washer...........Water purifier.................. Vacuum Cleaner............
G. MIxi.......Electric automatic Press................Grinder............Sewing Machine............Oven..................Toaster...............
H. Camera.............Tape-recorder....................Handycam..............

13. What type of Newspaper/magazine do You Purchase? If They are more than one tick also

A. Daily ( )
B. Weekly ( )
C. Monthly ( )
D. Quarterly ( )
E. Sometimes ( )
F. Never ( )

14. What is the Average Expenditure on Newspaper/magazine?

A. Rs. 100 ( )
B. Rs. 200 ( )
C. Rs. 300 ( )
D. Rs. 500 ( )
E. Above Rs. 500 ( )

15. Where do You Like to go for Shopping?

A. Any/ordinary shop ( )
B. Branded showrooms ( )
C. Mall ( )
D. Halt ( )

16. What is the Opinion of Your Friends/other People about You and Your Family?

A. Most prosperous Family ( )
B. Prosperous Family ( )
C. Middle Class ( )
D. Average Class ( )
E. Low Class ( )

17. What should be the basis of Social Status according to the Members of Your Family?

A. According to profession and work ( )
B. On the basis of property and economic condition ( )
C. According to caste ( )

PART-B

1. **Name of the respondent** ....
2. **Sex M**................................ **F**..
3. **Age**..
4. **Place**..
5. **Do you have any knee problem?**

   Yes...........................

   No............................

6. **If yes than from how long?**

   A. Last 6 months – 12 months ( )
   B. Last 12 months – 18 months ( )
   C. Last 18 months – 24 months ( )

D. More than 24 months ( )

7. When You Feel more Pain?

A. During standing ( )
B. During sitting ( )
C. During walking ( )
D. All the time ( )

8. Are You able to sit at the Floor?

Yes...............................

No................................

9. Which Type of Toilet do You have in Your Home?

A. Indian ( )
B. European ( )
C. Both ( )

10. Which Type of Toilet You Prefer to Use?

A. Indian ( )
B. European ( )
C. Both ( )

11. Do You have any Problem while using Indian Toilet?

Yes...................

No.....................

12. Which type of Problem?

A. Pain in knee joints ( )
B. Numbness in feet ( )
C. Numbness in calf muscles ( )
D. Standing after sitting ( )

13. How You fall in Pain in Knees, after?

A. Fall ( )
B. Slip ( )
C. Accident ( )
D. Fracture ( )

14. What was Your Age when this Problem Starts?

A. Before 30 ( )
B. 30 + ( )
C. 40 + ( )
D. 50 + ( )

15. Which type of Treatment You are taking for Cure

A. Medicine ( )
B. Physiotherapy ( )
C. Exercise/yoga ( )
D. Avoidance of strain ( )

16. Do You feel any Betterment?

A. Better with treatment ( )
B. Better after treatment ( )
C. Better with physiotherapy ( )
D. Better after physiotherapy ( )

17. Do You have any Fixture in the Toilet?

A. Handless at one side of the wall ( )
B. Handless at both the sides of the wall ( )
C. Any other ( )
D. None ( )

18. Which Type of Surface do you have in Your Toilet?

A. Raised platform of one step ( )
B. More than one stairs ( )
C. In the level ( )

19. Which Type of Surface of Floor do You have in Your Toilet?

A. Simple floor ( )
B. Marble floor ( )
C. Stone ( )
D. Tiles ( )

20. Which Type of Material of Door do You have in Your Toilet?

A. Wooden door ( )
B. Metal door ( )
C. Plywood door ( )

21. Which Type of Material of Windows do You have in Your Toilet?

A. Wooden ( )
B. Metal ( )
C. Glass ( )

22. Which Type of Surface of wall do You have in Your Toilet?

A. Painted ( )
B. Whitewashed ( )
C. Tiles ( )
D. Wall paper ( )

23. What Type of Material do You have in Your Toilet?

A. Plastic ( )
B. Wooden ( )
C. Metal ( )
D. Any other ( )

24. Which Type of Lighting do You have in Your toilet?

A. Bulb ( )
B. Fluorescent tube ( )
C. CFL ( )
D. Combination of bulb/flourescent tube/CFL ( )

25. If You don't have European Toilet in Your Home then why?

A. Financial problem ( )
B. Family problem ( )
C. Facility problem ( )
D. Don't like ( )

26. If You are provided then will You use that?

Yes.....................

No.......................

27. If no then why?

A. Do not like ( )
B. Do not have need ( )
C. Any other ( )

28. Have You met with any Accident in the Toilet like:

A. Fall due to having problem in standing after using Indian toilet ( )
B. Fall due to not having any support at the side/front of the wall ( )
C. Any other ( )

PART-C

1. **Weight of the Respondent ....**

2. **Anthropometric Measurements in sitting position**

S. No. Measurements	*Cm.*	
1. Sitting height		
2. Sitting shoulder height		
3. Buttock Popliteal length		
4. Popliteal height		
5. Elbow rest height		
6. Hip Breadth		
7. Palm length		
8. Grip strength	Right hand	Left hand

VALIDATION SHEET

Name of the respondent:

Statements	*Highly Satisfied*	*Satisfied*	*Not satisfied*
Score	3	2	1

Are you satisfied with the following statements?

	HS	S	NS
1. Height of chair			
2. Weight of chair			
3. Width of chair			
4. Depth of chair			
5. Height of hand rest			
6. Supporting handle			
7. Operation of assembling			
8. Removal of pan			
9. Able to carry chair			
10. Material of toilet chair			
11. Material of seat plate			
12. Material of back rest			
13. Material of hand rest			
14. Comfortableness			
15. Cost of chair			

Bibliography

Allaire, S.H.; La Valley, M.P.; Evans, S.R.; O'Connor, G.T.; Kelly-Hayes, M.; Meenan, R.F.; Levy, D. and Felson, D.T. (1999). Evidence for decline in disability and improved health among persons aged 55 to 70 years: *The Framingham heart study*. *Am. Public Health*, 89 (11): 1678-1683.

Anderson, C.; Ni Mhurchu, C.; Brown, P.M. and Carter, K. (2002). Stroke rehabilitation services to accelerate hospital discharge and provide home based care: *An Overview and Cost Analysis*. *Pharmacoecononics*, 20 (8): 537-552.

Anderson, R. and Lagerlof, E. (1983). Accident data in the new Swedish information system on occupational injuries. *Ergonomics*. 26: 33-42.

Armour, D.C. (2002). *Medicines in the elderly*. Pharmaceutical Press: London.

Ashworth, J.B.; Reuben, D.B. and Benton, L.A. (1994). Functional profile of healthy older person. *Age and Ageing*, 23: 34-39.

Aurelio, D.N. (2000). The experimenter/participant relationship: minimizing stress in applied research settings. In: *Proceedings of the International Ergonomics Association Conference*, San Diego, Canada, CD ROM.

Avlund, K. and Damsgaard, M.T. (2002). Tiredness in daily activities among non-disabled old people. *J. Clin. Epidemiol*. 55 (10): 965-973.

Avlund, K.; Damsgaard, M.T. and Schroll, M. (2001). Tiredness as determinant of subsequent use of health and social services among non-disabled elderly people. *J. Ageing Health*, 13 (2): 267-286.

Bales, C.W. and Ritchie, C.S. (2002). Sarcopenia, weight loss, and nutritional frailty in the elderly. *Ann. Rev. Nutr.*, 22: 309-323.

Barbaccia, J.C. (1995). Changes in physical function and the ageing process : implications for facility design. In D. Driver (ed.). *Proceedings of Blueprint for Ageing*. CA, Oakland.

Bayer, A.H. and Harper, L. (2000). Fixing to stay: A National Survey on Housing and Home Modification Issues. *Executive Summary*, Washington, DC.: *AARP*.

Blazer, D.G. (2002). Self-efficacy and depression in late life : A primary prevention proposal. *Ageing Ment. Health*, 6 (41): 315-324.

Blazer, D.G. (2003). Depression in late life : Review and commentary. *J. Gerontol A. Biol. Sci. Med. Sci.,* 58 (3): 249-265.

Blumenthal, J.; Babyak, M.; Moore, K.; Craighead, W.; Herman, W.; Khatri, P.; Waugh, R.; Napolitana, M.; Foman, L.; Applebacum, M.; Doraiswamy, P. and Krishnan, K. (1999). Effects of exercise training on older patients with major depression. *Arch. Intern. Med.*, 159 (19): 2349-2356.

Caplan, L.J. and Schooler, C. (2003). The role of fatalism, self-confidence and intellectual resources in the disablement process in older adults. *Psychol. Ageing*, 18 (3): 551-561.

Carmelli, D.; Kelly-Hayes, M.; Wolf, P.A. and Swan, G.E. (2000). The contribution of genetic influences to measures of lower extremity function in older male twins. *J. Gerontol A. Biol. Sci. Med. Sci.*, 55 (1): M849-853.

Chakrabarti, D. (1997). Ergonomics of the design of products, equipments and accessories. *Indian Journal of Indus Med.,* 20 (2): 49-63.

Chang, W.R.; Gronqvist, R. and Courtney, T.K. (2001c). The role of friction in the measurement of slipperiness. *Ergonomics*, 44: 1233-1261.

Chang, W.R.; Gronqvist, R. and Thorpe, S. (2001b). The role of friction in the measurement of slipperiness, *Ergonomics,* 44: 1217-1232.

Chang, W.R.; Kim, L.J.; Manning, D. and Bunterngchit, Y. (2001a). The role of surface roughness in the measurement of slipperiness. *Ergonomics,* 44: 1200-1216.

Christensen, K.; McGue, M.; Yashin, A.; Lachine, I.; Holm, N.V. and Vaupel, J.W. (2000).Genetic and environmental influences on functional abilities in Danish twins aged 75 years and older. *J. Gerontol A. Biol. Sci. Med. Sci.*, 55 (8): M446-452.

Clark, M.C.; Czaja, S.J. and Weber, R.A. (1990). Older adults and daily living task profiles. *Human Factors,* 32 (5): 537-549.

Coleman, R. (1997). Breaking the age barrier, speech at the Royal Society of Arts, June 97 as part of the Design Council – Design in Education Week, London: Design Age, Royal College of Art, March 12th. *http://valley.interact.nl/DAN/NEWSLETTER/NEWS97/home.html, 6, 6, 1998.*

Courtney, T.K. and Webster, B.S. (1999). Disability occupational morbidity in the United States: An alternative way of seeing the Bureau of labour statistics data. *Journal of Occupational and Environmental Medicine,* 41: 60-69.

Courtney, T.K.; Sorock, G.S. and Manning, D.P. (2001). Occupational slip, trip and fall related injuries can the contribution of slipperiness be isolated? *Ergonomics,* 44: 1118-1137.

Crassley, L. (2004). Bridging the emotional gap. In: Mc Donagh, D.; Hekkert, P.; Van Erp. J.; Gyi, D. (eds.). Design and emotion: The experience of Everyday Things. Taylor and Francis, London, pp. 37-42.

Demirbilek, O. and Demirkan, H. (1998). Involving the elderly in the design process. *Archit. Sci. Rev.*, 41 (4): 157-163.

Duncan, P.W.; Chandler, J.; Studensky, S.; Hughes, M. and Prectt, B. (1993). How do physiological components of balance affect mobility in elderly men? *Arch. Phys. Med. Rehabil.*, 74 (12): 1343-9.

Ebrahim, S. (2002). Ageing, health and society. *Int. J. Epidemiol.*, 31 (4): 715-8.

EUFRP, (1997). PRESENCE Project, European Union funded Research Programme, *http://www.presenceweb.org, 10, 3, 1998* .

Evans, W.A.; Courtney, A.J. and Fak, K.F. (1988). The design of school furniture for Hong-Kong school children: an anthropometrics case study. *Applied Ergonomics,* 19: 122-134.

Femia, E.E.; Zarit, S.H. and Johanson, B. (1997). Predicting change in activities of daily living : A longitudinal study of the oldest old in Sweden. *J. Gerontol, B. Psychol. Sci. Soc. Sci.*, 52 (16): 294-302.

Fingurhut, L.A.; Cox, C.S. and Warner, M. (1998). International comparative analysis of injury mortality. Findings from the ICE on injury statistics, Advance data from Vital and Health Statistics No. 303.

Freedman, V.A.; Martin, L.G. and Schoeni, R.F. (2002). Recent trends in disability and functioning among older adults in the United States, a systematic review. *JAMA,* 288 (24): 3137-3146.

Fulton Suri, J. and IDEO (2005). *Thoughtless Acts.* Chronicle Books, San Francisco.

Gill, T.M.; Williams, C.S. and Tinetti, M.E. (2000). Environmental hazards and the risk of non-syncopal falls in the homes of community living older persons. *Med. Care,* 38 (12): 1174-1183.

Giverchi, R. and Velasquez, V. (2004). Positive space. In: Mc Donagh, D.; Hekkert, P.; Van Erp. J.; Gyi, D. (eds.). *Design and emotion: The experience of Everyday Things.* Taylor and Francis, London, pp. 43-47.

Goldberg, T.H. and Chavin, S.I. (1997). Preventive medicine and screening older adults. *J. Am. Geriats Soc.*, 45 (3): 344-354.

Grandjean, E. (1980). *"Fitting the Task to the Man : An ergonomic Approach".* Taylor and Francis, London.

Green, W.S.; Kanis, H. and Vermeeren, A.P.O.S. (1997). Tuning the design of everyday products to cognitive and physical activities of users. In *Contemporary Ergonomics.* Taylor and Francis, London.

Grieve, D. and Pheasant, S. (1992). Biomechanics. In *"the Body at Work"*, edited by W.T. Singleton, Cambridge University Press, Cambridge, England.

Griffin, S. (1998). Diabetes care in general practice : Meta-analysis of randomized controlled trails. *BMJ,* 317 (7155): 390-406.

Grimley, Evans J. (2000). Ageing and medicine. *J. Intern. Med.*, 24 (2) : 159-167.

Gronqvist, R. (1999). Slips and falls in S. Kumar (ed). *Biomechanics in Ergonomics* (London: Taylor and Francis), 315-375.

Gronqvist, R. Abeysekera, J.; Gard, G. and Hsiang, S.M. (2001). Human centred approaches in slipperiness measurement. *Ergonomics,* 44: 1167-1199.

Gronqvist, R. and Roine, J. (1993). Serious occupational accidents caused by slipping. *Advances in Industrial Ergonomics and Safety,* 515-519.

Grover, R. 2002. Suitability of ergonomically designed table and chair for physically challenged children. Unpublished, Master Thesis, G.B.P.U.A.&T., Pantnagar.

Guralnik, J.M.; Ferrucci, L. and Balfour, J.L. (2001). Progressive versus catastrophic loss of the ability to walk. *J. Am. Geriatr. Soc.*, 49 (11): 1463-1470.

Gurnack, A.M. (1997). *Older adult's misuse of alcohol, medicines and other drugs,* Springer Publishing Company, New York.

Gyi, D.; Porter, J.M. and Case, K. (2000). Design practice and designing for all. In : *Proceeding of the International Ergonomics Association Conference,* San Diego, Canada, CD ROM.

Hagberg, M.; Christian, D.; Courtney, T.K. and Smith, T.J. (1997). Conceptual and definitional issues in occupational injur epidemiology. *American Journal of Industrial Medicine*, 32: 106-115.

Handoll, H.H.; Parker, M.J. and Sherrington, C. (2003). Mobilization strategies after hip fracture surgery in adults. *Cochrane Database Syst. Rev.*, 2003 (1): CD01704.

Harris, C.; Straker, L.; Pollock, C. and Trinidad, S. (2005). Musculo-skeletal outcomes in children using information technology the need for a specific etiological model. *International Journal of Industrial Ergonomics*, 35: 131-138.

Heikkinen, E. (2003). *What are the main risk factors for disability in old age and how can disability be presented?* WHO regional office for Europe's health evidence network synthesis report. Available online at *http://www.euro.int/document/hen/disabilities 0903.pdf.*

Heikkinen, E.; Heikkinen, R.L. and Ruoppila, I. (1997). Functional capacity and health of elderly people the evergreen project. *Scandinavian Journal of Social Medicine, Supplement*, 53.

Heikkinen, R.L. and Kauppinen, M. (2004). Depressive symptoms in late life : A 10 year follow-up. *Arch. Gerontol Geriatr.*, 38 (3): In press.

Heikkinen, R.L.; Berg, S.; Avlund, K. and Tormakangas, T. (2002). Depressed mood : Changes during a five years follow-up in 75 years old men and women in three Nordic localities. *Ageing Clin. Exp. Res.*, 14 (3 Sup.): 16-28.

Helander, M. and Nagamichi, M. (eds.) (1992). *Design for manufacturability : A system approach to concurrent engineering and ergonomics.* Taylor and Francis, London.

Hirvensalo, M.; Rantanen, T. and Heikkinen, E. (2000). Mobility difficulties and physical activity as predictors of mortality and loss of independence in the community living older population. *J. Am. Geriatr. Soc.*, 48 (5): 493-508.

Hsiao, E.T. and Robinovitch, S.N. (1998). Common protective movements govern unexpected falls from standing height. *Journal of Biochemics*, 31: 1-9.

Hutchins, L.F.; Unger, J.M.; Crowley, J.J.; Coltman, C.A. and Albain, K.S. (1999). Under representation of patients 65 years of age or older in cancer treatment trails. *N. Engl. J. Med.*, 341 (27): 2061-2067.

International Conference on Harmonization of Technical Requirements of Pharmaceuticals for Human Use (1993). Studies in support of special populations: geriatrics : ICH harmonized tripartite guideline, E7. Available online at: *http: / / www. ich. org / Mediaserver. jser? @_ID = 4834 @-MODE=GLB.*

Jeong, B.Y. and Park, K.S. (1990). Sex differences in anthropometry for school furniture design. *Ergonomic*, 33: 1511-1521.

Johnson, J.; Weissman, M.M. and Klesman, G.L. (1992). Service utilization and social morbidity associated with depressive symptoms in the community. *JAMA*, 267 (11): 1478-1783.

Joliffe, J.A.; Rees, K.; Taylor, R.S.; Thompson, D.; Oldridge, N. and Ebrahim, S. (2002). Exercise based rehabilitation for coronary heart disease (*Cochrane review).* I: Cochrane Library: 4, Oxford, Update Software.

Jordan, P.W. and Thomas, B. (1995). But how much extra would you pay for it? *In Contemporary Ergonomics* (1995), eds. S. Robertson, pp. 145-148. Taylor and Francis, London.

Jordan, P.W.; Thomas, B. and McClelland, I.I. (1996). *Issues for usability evaluation in industry*: seminar discussion, pp. 237-243, Taylor and Francis, London.

Judd, L.L.; Paulus, M.P.; Wells, K.B. and Rapaport, M.H. (1996). Socio-economic burden of subsyndramal depressive symptoms and major depression in a sample of the general population. *Am. J. Psychiatry*, 153 (11): 1411-7.

Kanis, H. (1997). Variation in measurement repetition of human characteristics and activities. *Applied Ergonomics*, 28 : 155-163.

Kanis, Hand and Vermeeren, A.P.O.S. (1996). Teaching user involved design in the delft curriculum. *In Contemporary Ergonomics*, Taylor and Francis, London, pp. 374-378.

Kayis, B. (1991). Why ergonomie designs in school, in V. Popavic and M. Walkar (eds.). *Proceedings of the 27th Annual Conference of the Ergonomic Society of Australia,* Ergonomic and Human Environments, Column.

Kemmlert, K. and Lundholm, L. (2001). Slip, trip and falls in different work groups with reference to age, safety science, 28: 59-75.

King, A.; Rejeski, W.J. and Buchner, D.M. (1998). Physical activity intervention targeting older adults: a critical review and review recommendations. *Am. J. Prev. Med.*, 15 (4): 316-333.

Kinsella, K. and Velkoff, V.A. (2001). *An aging world*: 201, U.S. Census Bureau, series. p 95/01-1, U.S. Government Printing Office, Washington, DC.

Kreifeldt, J.G. (1991). "Application of ergopnomics in product designing". Proc. International Symposium on Ergonomics, Occupational Health Safety and environment, held at 1st December, IIT, Bombay, India, in January 1991.

La Croix, A.Z.; Guralnik, J.M.; Berkman, L.F. Wallace, R.B. and Satterfield, S. (1993). Maintaining mobility in late life. II. Smoking, alcohol consumption, physical activity and body mass index. *Am. J. Epidemiol.*, 137 (8): 858-869.

Laflamme, L. and Menckel, E. (1995). Ageing and occupational accidents: a review of the literature of the last three decades. *Safety Sci.*, 21: 1445-161.

Laukkanen, P.; Kauppinen, M.; Era, P. and Keikkine, E. (1993). Factors related to coping with physical and instrumental activities of daily living among people born in 1904-1923. *Int. J. Gerriatr. Psychiotry*, 8: 287-296.

Laukkanen, P.; Keikkinen, E.; Kauppinen, M. and Kallinen, M. (1992). Use of drugs by non-institutionalized urban Finns born in 1904-1923 and the association of drug use with mood and self-rated health. *Age Ageing*, 21 (5): 343-352.

Laukkanen, P.; Leskinen, E.; Kauppinen, M.; Sakari-Rantala, R. and Heikkinen, E. (2000). Health and functional capacity as predictors of community dwelling among elderly people. *J. Clin. Epidemiol.*, 53 (9): 257-265.

Lawton, M.P. (1980). *Environmental and ageing*. Brooks/ Cole Pub. Co. Monterey, California.

Leaman, T.B. and Murphy, P.L. (1995). Occupational slips and falls: More than a trivial problem. *Ergonomics,* 38: 487-498.

Leclercq, S. (1999). The prevention of slipping accidents : a review and discussion of work related to the methodology of measuring slip resistance. *Safety Sci.,* 31: 95-125.

Lee, P.Y.; Alexander, K.P.; Hammil, B.G.; Pasquali, S.K. and Peterson, E.D. (2001). Representation of elderly persons and women in published randomized trails of acute coronary syndromes. *J. Am. Med. Assoc.*, 286 (6): 708-713.

Leveille, S.G.; Guralnik, J.M.; Ferrucci, L. and Langlois, J.A. (1999). Ageing successfully until death in old age: opportunities for increasing active life expectancy. *Am. J. Public Epidemiol.*, 149 (7): 654-664.

Linden, M.; Horhas, A.L.; Gilberg, R. and Steinhagen-Thiessen, E. (1997). Predicting health care utilization in the very old. The role of physical health, mental health, attitudinal and social factors. *J. Ageing Health,* 9 (1): 3-27.

Lloyd, D.G. and Stevenson, M.G. (1992). *Investigation of floor surface profile characteristics that will reduce the incidence of slips and falls*. Mechanical Engineering Transaction, Institute of Engineering (Australia), 17 (2): 99-104.

Lloyd-Williams, F.; Mair, F.S. and Leither, M. (2002). Exercise training and heart failure : A systematic review of current evidence. *Br. J. Gen. Pract.*, 52 (474): 47-55.

Lorig, K.R.; Sobel, D.S.; Stewart, A.L.; Brown, B.W. Jr.; Bandura, A.; Ritter, P.; Gonzallez, V.M.; Laurent, D.D. and Holman, H.R. (1999). Evidence suggesting that chronic disease self-management programme can improve health status while reducing hospitalization: A randomized trail. *Med. Care*, 37 (1): 5-14.

Maida, J.C. (1993). *Review of Jack* 5, 4. "Ergonomics in Design". July: 35-36.

Mandal, A.C. (1982). The correct height of school furniture. *Human factors*, 24: 257-269.

Manton, K.G.; Corder, L.S. and Stallard, E. (1993). Estimates of change in chronic disability and institutional incidence and prevalence rates in the U.S. elderly population from the 1982, 1984 and 1989, National Long Term Care Survey. *J. Gerantol. Soc. Sci.*, 48 (4): 5153-5160.

Marshall, R.; Case, K.; Gyi, D.; Oliver, R. and Porter, M. (2002). A task based 'design for all' support tool. *Journal of Robotics and Computer Integrated Manufacturing*, 18: 297-303.

Matsubara, Y. and Nagamachi, M. (1997). Hybrid Kansei engineering system and design support. *Int. J. Ind. Ergon.*, 19: 81-92.

Maudi, A.M. and Choobinch, A.R. (1997). Static anthropometric characteristics of students age range size 11 in Mazandaran province/Iron and school furniture design based on ergonomics principles. *Applied Ergonomics*, 28 (2): 145-147.

McElbinney, J.; Koval, K.J. and Zucherman, J.D. (1998). Falls in the elderly, *Archives of the American Academy of Orthopaedic Surgeons.* 2 (1): 50-55.

Mehr, D.R. and Tatum, P.E. (2002). Primary prevention of diseases of old age. *Clin Geriatr. Med.*, 18 (3): 407-430.

Mital, A. (1994). Issues and concerns in accommodating the elderly in the workplace. *Journal of Occupational Rehabilitation,* 4: 253-267.

Munck-Ulfrfalt, U.; Falek, A.; Forsberg, A.; Dahlin, C. and Eriksson, A. (2003). Corporate ergonomics programme at Volvo car corporation. *Appl. Ergon.*, 34 (1): 17-22.

Murphy, S.L. (2000) *Deaths: Final data for 1998*, National vital statistics reported, Vol. 48, no. 11 Hyattsville (MD): National Centre for Health Statistics.

Nagamachi, M. (1995). Kansei Engineering: A new ergonomic consumer oriented technology for product development. *Int. J. Ind. Ergon.,* 15: 3-11.

Nagamachi, M. (2002). Kansei engineering as a powerful consumer oriented technology for product development. *Appl. Ergon.,* 33: 289-294.

National Safety Council, (1995). *International Accident facts* (Itasca, IL: National Safety Council).

Nilsson, S.E.; Johansson, B.; Berg, S.; Karlsson, D. and McClearn, G.E. (2002). A comparison of diagnosis capture from medical records, self reports, and drug registrations: A study in individuals 80 years and older. *Ageing Clin. Exp. Res.*, 14 (3): 178-84.

Norris, B.J. and Wilson, J.K. (1995). CHILTDATA : *The handbook of child measurements and capabilities : Data for design safety,* Department of Trade and Industry, London, U.K.

Norris, B.J. and Wilson, J.R. (1997). *Designing safety into products making ergonomics evaluation a part of the design process.* The University of Nottingham, Nottingham, U.K.

Office for National Statistics (1998). *Social Trends.* 28th Edition NMSO, London.

Oliver, R.; Gyi, D.; Porter, M.; Marshall, R. and Case, K. (2001). A survey of the design needs of older and disabled people. In: *Contemporary Ergonomics.* McCabe, P.T.; Hanson, M.A. and Robertson, S.A. (Eds.). Taylor and Francis, London. pp. 365-370.

Peebles, L. and Norris, B.J. (1998). ADULTDATA. The *handbook of adult anthropometric and strength measurements data for design safety*. Department of Trade and Industry, London, U.K.

Penninx, B.W.; Guralnik, J.M.; Ferrucci, L. : Simonsick, E.N.; Deeg, D.J. and Wallace, R.B. (1998). Depressive symptoms and physical decline in community dwelling older persons. *JAMA*, 279 (21): 1720-6.

Pheasant, S. (2001). *Body space*. Taylor and Francis, London.

Porter, J.M.; Case, K. and Freer, M.T. (1999). Computer aided design and human models. In: Karwowaski, W. (eds.). *Handbook of Occupational Ergonomics*. CRC Press LLC, Florida, pp 479-500.

Porter, J.M.; Case, K.; Marshall, R.; Gyi, D.E. and Sims, R.E. (2004). Beyond Jack and Jill : Designing for individuals within populations using HADRIAN. *International Journal of Industrial Ergonomics*, 33: 249-264.

Rantanen, J. (1999). Research challenges arising from changes in work life, Scandinavia. *Journal of Work, Environment and Health*, 25 (6): 473-483.

Rantanen, T. and Avela, J. (1997). Leg extension power and walking speed in very old people living independently. *J. Gerontol A. Biol. Sci. Med. Sci.*, 52 (4): 225-231.

Rantanen, T.; Guralnik, J.M.; Foley, D.; Masaki, K. Leveille, S.; Curb, J.D. and White, L. (1999b). Middle handgrip strength as a predictor of old age disability. *JAMA*, 281 (6): 558-560.

Ranzijn, R. (2002). The potential of older adults to enhance community quality of life. *Ageing International*, 27 (2): 30-35.

Redfern, N.S.; Cham, R. and Powers, C. (2001). Biomechanics of slips. *Ergonomics*, 44: 1138-1166.

Rice, D.P. and Laplante, M.P. (1992). Medical expenditures for disability and disabling co-morbidity. *Am. J. Public Health,* 82 (5) : 739-741.

Richardson, S. and Poulson, D. 1996. Usability–disability and elderly. *3rd Annual Consultation meeting of the Telematics applications programmes*, applications for the European Information Society. Brussels: Palais Des Congress, Dec. 2-3.

Robertson, A. and Tracy, C.S. (1998). Health and productivity of older workers, Scandinavia. *Journal of Work, Environment and Health*, 24 (2): 85-97.

Rochon, P.A.; Berger, P.B. and Gordon, M. (1998). The evolution of clinical trails: inclusion and representation. *Can Med. Assoc. J.,* 159 (11): 1373-1374.

Rybarczyk, B.; DeMarco, G.; Dela Cruz, M. and Lapidos, S. (1999). Comparing mind body wellness interventions for older adults with chronic illness: Classroom versus home instruction. *Behav. Med.*, 24 (4): 181-190.

Sadena, M.R. and Glass, P.S. (2003). *Pharmacokinetics in the elderly.* Best practice and research clinical anesthesiology, 17: 191-205.

Sahyoun, N.R. (2002). Nutrition education for the healthy elderly population: isn't it time? *J Nutr. Educ. Bahv.*, 34 (Suppl. I): 542-7.

Salive, M.E. and Guralnik, J.M. (1997). Disability outcomes of chronic disease and their implications for public health. *Public Health and Ageing*, pp. 87-106.

Sandhu, J.S. (2001). An integrated approach to universal design toward the inclusion of all ages, cultures and diversity. *Universal Design Handbook*, New York, 303-314.

Sattin, R.W. (1992). Falls among older persons: A public health perspective. *Annual Review of Public Health,* 13 : 489-508.

Scott, M.A.C.; Nowlan, S. and Gutman, G. (2001). Progressive housing design and home technologies in Canada. *Universal Design Handbook,* New York, pp 36.1-36.15.

Shumway, Cook, A.; Patla, A.; Stewart, A.; Ferrucci, L.; Ciol, M.A. and Guralnik, J.M. (2003). Environmental components of mobility disability in community-living older persons. *J. Am. Geriatr. Soc.*, 51 (3): 393-408.

Singh and Saxena, 1981. Socio-economic scale for rural urban group. *Psychological Research Cell*, Agra.

Sletvold, O.; Tilvir, R.; Jonsson, A.; Schroll, M. and Snaedal, J. (1996). Geriatric workup in the Nordic countries. The Nordic approach to comprehensive geriatric assessment. *Dan. Med. Bull.*, 43 (4): 350-409.

Smith, D.B.; Meshkati, N. and Robertson, M.M. (1993). The older driver and passenger. In: Peacock, B., Karwowaski. *Automotive Ergonomics*. Taylor and Francis, 453.479.

Smith, S.A.; Norris, B.J. and Peebles, L. (2000). *Older adultdata*—The *handbook of measurements and capabilities of the older Adult* –Data for Design safety. Department of Trade and Industry, London, U.K.

Smits, C.H.M.; Deeg, D.J.H. and Jonker, C. (1997). Cognitive and emotional predictors of disablement in older adults. *J. Ageing Health,* 9 (2): 204-221.

Snedecor, G.W. and Cochran, W.G. 1967. *Statistical Methods*. 6th edition, Lowa State University Press, Ames, Lowa, p. 558.

Sonn, U.; Grimby, G. and Svanborg, A. (1996). Activities of daily living studied longitudinally between 70 and 76 years of age. *Disabil. Rehabil.*, 18 (2): 91-100.

Steinfeld, E. and Danfort, S. (1993). *Automated doors: towards universal design.* IDEA Publications, SUNNY/Buffalo: The centre for Inclusive Design and

Environmental Access. *http://www.adequatev.org/~idea/publications/papers/autodoor_art.html, 11, 3: 1977.*

Steinfeld, E. and Shea, S.M. (2001). *Enabling home environments. Strategies for ageing in place.*

Stewart, A.L.; Mills, K.M.; Sepris, P.G.; King, A.C.; Mc Lellan, B.Y.; Roitz, K. and Ritter, P.L. (1997). Evaluation of CHAMPS, a physical activity promotion programme for older adults. *Ann. Behav. Med.*, 19 (4): 353-361.

Story, M.F.; Mueller, J.L. and Mace, R.L. (1998). The universal design, designing for people of all ages and abilities. *The Centre for Universal Design, NC.*

Strandberg, L. (1985). The effect of conditions underfoot on falling and overexertion accidents. *Ergonomics*, 28: 131-147.

Sulander, T.T.; Rahkonen, O.J. and Uutela, A.K. (2003). Functional ability in the elderly finish population: Time period differences and associations, 1985-99. *Scand. J. Public Health*, 31 (2): 100-106.

The Henley Centre and Design Age, (1997). The older marketplace a goldmine but still a grey area. Presentation by the Henley Centre and Design Age (*Royal College of Art*) at the Royal college of Art, London, March, 1997.

Tinetti, M.E.; Lnouge, S.K.; Gill, T.M. and Doucette, J.T. (1995). Shared risk factors for falls, incontinence and functional dependence. *JAMA*, 273 (17): 1348-1353.

Tsuji, I.; Kuwahara, A.; Nishino, Y.; Ohkuboto, T.; Sasaki, A. and Hisamichi, H. (1999). Medical cost for disability: A longitudinal observation of national health insurance beneficiaries in Japan. *J. Am. Geriatr. Soc.*, 47 (4): 470-506.

U.S. Bureau of the Census (1996). Population Projections of the United State by age, Sex, Race and Hispanic

Origin: 1995-2050, *Current Population Reports*. pp. 25-1130, U.S. Government Printing Office, Washington, DC.

United Nations Population Division (1999). World population projects. The 1998 Revision. *www.popin.org/pop1998*. United Nations, New York.

Van Wely, P. (1970). Design and disease. *Applied Ergonomics*, 1: 262-269.

Warner, M.; Barnes, P.M. and Fingerhut, L.A. (2000). Injury and poisoning episodes and conditions: National Health Interview Survey, 1997. *Vital and Health Statistics, Science* 10, No. 303.

Wells, K.B.; Stewart, A;. Hays, R.D.; Burnam, M.A.; Rogers, W. and Daiels, M. (1989). The functioning and well- being of depressed patients. Results from the Medical Outcomes study. *JAMA*, 262 (7): 914-9.

West, S.K.; Rubin, G.S.; Broman, A.T.; Munoz, B.; Bandeen-Roche, K. and Tura, K. (2002). How does visual impairment affect performance on task of everyday life? The SEE Project. Salisbury Eye Examination. *Arch. Opthalmol.*, 120 (6): 774-780.

Winblad, I.; Jaaskelainen, M.; KIvela, S.L.; Hiltunen, P. and Laippala, P. (2001). Prevalence of disability in three birth cohorts at old age over time spans of 10 and 20 years. *J. Clin. Epidemiol.*, 54 (10): 1019-1024.

Wolff, C. (1997). A haven, not a hospital: Assisted Living Design involves Practical and Emotional Issues. *Lodging Hospitality*. November 1st, 53 (3): 64.

Wolinsky, F.D.; Culler, S.D.; Callahan, C.M. and Johnson, R.J. (1994). Hospital resource consumption among older adults: a prospective analysis of episodes length of stay, and charges over a seven year period. *J. Gerontol. Soc. Sci.*, 49 (5): 5240-5250.

World Health Organization (1995). Physical status: The use an interpretation of anthropometry. Report of a

WHO expert committee. *Technical Report Series No. 854,* Geneva.

World Health Organization (2001). ICF International Classification of functioning, disability and health. Geneva: World Health Organization. Template available online at : *http://www.who.int/classification. icf.*

Index